DEVELOPMENT OF
MEMRISTOR BASED CIRCUITS

WORLD SCIENTIFIC SERIES ON NONLINEAR SCIENCE

Editor: Leon O. Chua
University of California, Berkeley

Series A.   MONOGRAPHS AND TREATISES*

*To view the complete list of the published volumes in the series, please visit:
http://www.worldscientific.com/series/wssnsa

WORLD SCIENTIFIC SERIES ON
NONLINEAR SCIENCE   Series A   Vol. 82

Series Editor: Leon O. Chua

DEVELOPMENT OF MEMRISTOR BASED CIRCUITS

Herbert Ho-Ching Iu
Andrew L Fitch
The University of Western Australia, Australia

World Scientific

NEW JERSEY · LONDON · SINGAPORE · BEIJING · SHANGHAI · HONG KONG · TAIPEI · CHENNAI

Published by

World Scientific Publishing Co. Pte. Ltd.
5 Toh Tuck Link, Singapore 596224
USA office: 27 Warren Street, Suite 401-402, Hackensack, NJ 07601
UK office: 57 Shelton Street, Covent Garden, London WC2H 9HE

British Library Cataloguing-in-Publication Data
A catalogue record for this book is available from the British Library.

World Scientific Series on Nonlinear Science, Series A — Vol. 82
DEVELOPMENT OF MEMRISTOR BASED CIRCUITS

ISBN 978-981-4383-38-7

Printed in Singapore.

Preface

Memristors are the missing fourth fundamental circuit element postulated by L.O. Chua in 1971. There was little research activity on this subject until researchers of Hewlett-Packard announced the solid state implementation of memristors in 2008. As memristors are not yet on the market, the development of memristor emulators and memristor based circuits is very important for real and practical engineering applications.

The objectives of this book are to review the basic concepts of the memristor, describe state-of-the-art memristor based circuits and to stimulate further research and development in the area. This book consists of eight chapters. Chapter 1 gives the introduction and description of some memristor emulators. In Chapter 2, a chaotic memristor based circuit is constructed, and a twin-T notch filter is used to control the observed chaotic behavior. In Chapter 3, hyperchaos in a memristor based modified canonical Chua's circuit is systematically studied. In Chapter 4, an analog memristor emulator based on a light dependent resistor is realized. Chapters 5 and 6 introduce memcapacitor emulators based on transformation of memristors. In Chapter 7, a study of chaos in a memristively coupled harmonic oscillators is presented. Finally, Chapter 8 gives the conclusion and future work.

The authors are indebted to the Editor of World Scientific Series on Nonlinear Science, Prof. Leon Chua, and to Senior Editor Ms. Lakshmi Narayanan for their help and congenial processing of the book.

Herbert Ho-Ching Iu
Andrew Fitch

Contents

Chapter 1

Introduction to Memristors

After the successful solid state implementation of memristors, a lot of attention has been drawn to the study of memristor based circuits and systems. In this chapter, basic concepts of memristor are reviewed. The HP solid-state memristor is introduced as well. In particular, some existing memristor simulators are discussed.

1.1 Introduction

Memristors have been well known as the missing fourth element postulated by L.O. Chua in 1971 [1] until researchers in Hewlett-Packard announced a solid state implementation of memristors in 2008 [2].

In circuit theory, the three basic two-terminal circuit elements are defined in terms of a relationship between two of the four fundamental circuit variables, namely, the *current i*, the *voltage v*, the *charge q* and the *flux φ*. There are six possible combinations of these four variables, five of them are well known. Two of them are given by

$$q(t) = \int_{-\infty}^{t} i(\tau)d\tau \tag{1.1}$$

and

$$\varphi(t) = \int_{-\infty}^{t} v(\tau)d\tau. \tag{1.2}$$

Three other relationships are given by definitions of the three classical circuit elements, namely, the *resistor*, the *inductor* and the *capacitor*,

$$dv = R \, di, \quad d\varphi = L \, di, \quad dq = C \, dv. \tag{1.3}$$

The relationship between the *flux* and the *charge* remains undefined. In 1971, Chua postulated the existence of a fourth basic two-terminal circuit element that is characterized by a φ versus q constitutive relation [1]. This element is called memristor. The symbol of the memristor and a hypothetical φ-q curve are shown in Fig. 1.1. The four pairs of circuit variables defining the four basic circuit elements are described in Fig. 1.2.

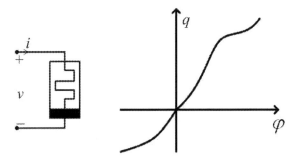

Fig. 1.1. Symbol of memristor and the φ-q curve.

A memristor is characterized by a constitutive relation $f(\varphi, q) = 0$. The memristor is *charge-controlled* if this relation can be expressed as a single-valued function $\varphi = \varphi(q)$ of the charge q. The memristor is *flux-controlled* if this relation can be expressed as a single-valued function $q = q(\varphi)$ of the flux φ.

The voltage across a *charge-controlled* memristor is given by

$$v = M(q) \, i, \tag{1.4}$$

where

$$M(q) = \frac{d\varphi(q)}{dq}. \tag{1.5}$$

Similarly, the current of a *flux-controlled* memristor is given by

$$i = W(\varphi) \, v, \tag{1.6}$$

where

$$W(\varphi) = \frac{dq(\varphi)}{d\varphi}. \tag{1.7}$$

$M(q)$ and $W(\varphi)$, representing the slope of a scaler function $\varphi = \varphi\,(q)$ and $q = q(\varphi)$ respectively, are called the *memristance* and *memductance* respectively.

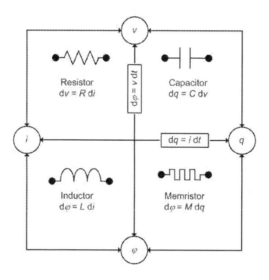

Fig. 1.2. The four basic circuit elements and their relationships.

The memristor is characterized by a relation between the charge q and the flux φ. This relation can be generalized to include any class of two-terminal device whose resistance depends on the internal state of the system [3]. These systems are called memristive systems.

Mathematically, an nth-order current-controlled memristive system is described by [3]

$$V_M(t) = M(x, I, t)I(t) \tag{1.8}$$

$$\dot{x} = f(x, I, t) \tag{1.9}$$

where x is a vector representing n internal state variables, $V_M(t)$ and $I(t)$ denote the voltage and current across the device, and M is the memristance.

Similarly, an nth-order voltage-controlled memristive system is described by [3]

$$I(t) = G(x, V_M, t)V_M(t) \qquad (1.10)$$

$$\dot{x} = f(x, V_M, t) \qquad (1.11)$$

where we call G is the memductance.

1.2 HP Memristor

The solid-state memristor developed by HP, shown in Fig. 1.3, is in the form of a partially doped titanium dioxide thin film with platinum electrodes [2]. In this model, by analyzing how the moving process influences the memristance value, the following results were obtained:

$$v = \left(R_{ON}\frac{w}{D} + R_{OFF}(1 - \frac{w}{D}) \right) i, \qquad (1.12)$$

$$w = \mu_v \frac{R_{ON}}{D} q. \qquad (1.13)$$

Assuming $R_{ON} \ll R_{OFF}$, the memristance is:

$$M(q) = R_{OFF}(1 - \frac{\mu_v R_{ON}}{D^2} q) \qquad (1.14)$$

where D is the total width of the titanium dioxide film, w is the width of the region of high dopant concentration on the titanium dioxide film, and R_{OFF} and R_{ON} are the limit values of the memristor resistance for $w = 0$ and $w = D$. μ_v is the dopant mobility.

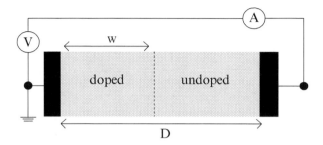

Fig. 1.3. The structure of HP memristor.

The characteristic element of the HP memristor is that when an AC voltage is applied to the device, the current versus voltage plot describes a pinched hysteresis loop as shown in Fig. 1.4.

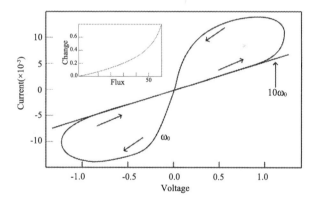

Fig. 1.4. The I-V characteristics of the HP memristor.

1.3 Memristor Emulator

Up to now, memristors are not yet available on the market due to the cost and technical difficulties in fabricating nano-scale devices. In order to study various applications of memristor, emulators of memristor have to be developed. In this subsection, we will briefly introduce several memristor emulators developed by researchers recently.

1.3.1 *Valsa's memristor emulator*

Valsa *et al.* [4] developed a flux controlled memristor modeling circuit in 2010. The circuit is intended to be used for obtaining proof of theoretical concepts regarding memristor properties by observing a memristive response to different input signals. The circuit uses a JFET to provide voltage controlled conductance (G_M) and an operational amplifier based integrator to obtain a voltage representing flux (v_ϕ) which is used to control G_M. In response to an input sinusoidal signal of the appropriate frequency and amplitude, the circuit enables the observation of a characteristic hysteretic loop passing through zero.

Due to the use of an integrator sub-circuit to obtain the voltage corresponding to flux, the circuit's operating region in terms of frequency is limited to signals lower than 5 Hz. This region can be adjusted to work with higher frequency input signals by changing the values of either the resistor R or capacitor C and thereby altering the RC relationship of the integrator, as seen in Fig. 1.5, but the range of the operating region will still remain at a similar width.

This circuit cannot be used to emulate a memristor in a functional circuit; it is designed for observing memristive behavior in an experimental environment rather than a practical one.

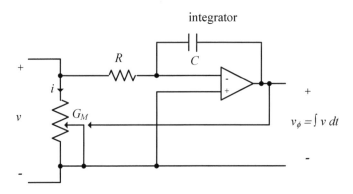

Fig. 1.5. Basic schematic of memristor modeling circuit [4].

1.3.2 *Pershin and Di Ventra's memristor emulator*

Pershin and Ventra [5] developed a programmable analog circuit, also presented in 2010. This circuit approached the design of a suitable memristor emulator from a completely different direction. The approach was taken based on the following concept. The thin film TiO_2 memristors behave as ideal memristors, whereas their behavior is not ideal and considerably more complex [5]. Pershin and Ventra chose to use pulse trains to set the memristor's state and then use low voltage analog signals for actual signal processing. This exploits the threshold behavior observed in memristors where larger signals alter the conductance of the device and much smaller signals can pass through with no or negligible effect [6, 7].

The block diagram, given in Fig. 1.6, shows that the circuit contains a microprocessor, an analog-to-digital converter (ADC) and a digital potentiometer. The circuit operates with the ADC converting the analog signal to its binary equivalent and supplying this to the microprocessor. The microprocessor then sets the digital potentiometer to the resistance value required. According to Pershin and Ventra [5], the microprocessor is coded to process signals according to

$$I(t) = R_M{}^{-1}(x, V_M, t)V_M(t) \tag{1.15}$$

$$\dot{x} = f(x, V_M, t) \tag{1.16}$$

where x is a vector representing n internal state variables, R_M is memory resistance and $R_M{}^{-1}$ is its inverse, being memductance [5]. In the work of Valsa *et al.* [4], it has been pointed out that the disadvantages of using a digital system to emulate an analog component are the limited resolution in the equivalent digital signal and the physical properties of the digital potentiometer. This comment can be justified considering the digital potentiometer used is a 256 position 10 kΩ type [5] which would give steps of 39 Ω. The circuit is also limited by the ADC sampling frequency of 1 kHz, meaning that the operating signals applied to the circuit should be at most 50 Hz and ideally less than 10 Hz [5].

Pershin and Ventra [5] have pointed out that the ADC, microprocessor and digital potentiometer can all be upgraded to provide greater frequency handling capability and stepping resolution, and that the components chosen for the demonstration are good enough to provide proof of theory. The advantages of their method may become more evident when further experimental data describing the operational characteristics and parameters of actual memristors are obtained and published. At that time, it is a relatively simple process to upgrade the basic system and alter the coding to suit the data as it is revealed. Furthermore, there will not be just one type of memristor; in the same sense there are a variety of capacitors in terms of construction materials, capacitance value and performance ratings. It can thus be expected that a number of different memristors and memristive systems will be developed in the near future.

Pershin and Ventra's circuit is an adaptable design that can be modified to effectively emulate the many variations in memristor types that are expected to be available.

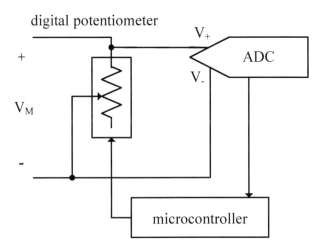

Fig. 1.6. Block diagram of memristor emulator [5].

1.3.3 *Muthuswamy's memristor modeling circuit*

The memristor modeling circuit presented by Muthuswamy and Chua [8] does not conform to the ideal memristor described by Chua in 1971 [1], nor does it follow the HP model [2]. It is more specifically a memristive device as described by Chua and Kang [3]. The equations describing the memristive system are given as [8]

$$V_M = \beta(x^2 - 1)i_M \qquad (1.17)$$

$$\dot{x} = i_M - \alpha x - i_M x \qquad (1.18)$$

where x represents the memristor's internal state, α and β are constants. The memristive symbols described by (1.17) and (1.18) are given in Fig. 1.7 and the plot derived from them is given in Fig. 1.8.

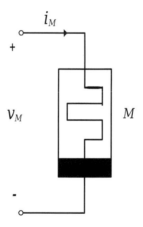

Fig. 1.7. Memristor circuit symbol [8].

The paper by Muthuswamy and Chua [8] describes the application of the memristor modeling circuit in a simple chaotic oscillator. For the chaotic oscillator circuit to operate correctly, it requires three independent state variables, which cannot occur in a single loop circuit with an ideal charge controlled type memristor.

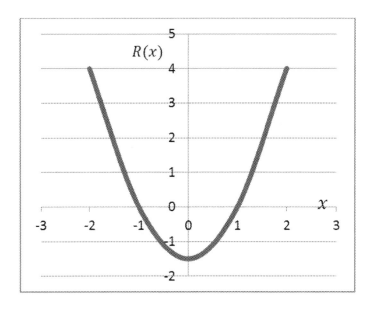

Fig. 1.8. Plot of the memristive function of the memristor modeling circuit [8].

1.3.4 *Kim's memristor emulating circuit*

In Kim's memristor emulating circuit [9], the input resistance is designed as a function of time integral of the input current. The equation governing the voltage and current of the memristor emulator is given as

$$v_{in} = \left(R_s + \frac{q_c}{C} \times R_T\right) i_{in} \tag{1.19}$$

where the symbols are defined in Figs. 1.9 and 1.10.

It is noted that the memristance of this model consists of a fixed part R_s and a variable part $(q_c/C) \times R_T$. Various features of the emulator are tested via circuit measurements and SPICE simulations, and it is found that the emulator behaves similarly as the real HP TiO_2 memristor.

In addition, Kim's memristor emulator is expandable, the emulator can be connected in serial, in parallel, or in hybrid (serial and parallel combined) with other memristors with identical or opposite polarity. The device is useful for developing other memristor applications [10, 11].

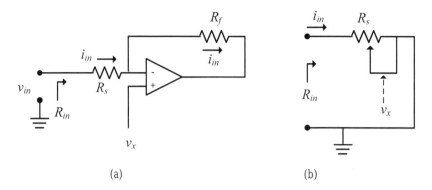

Fig. 1.9. Kim's concept of memristor emulator [9]. (a) Input resistance as a function of v_x. (b) Equivalent circuit.

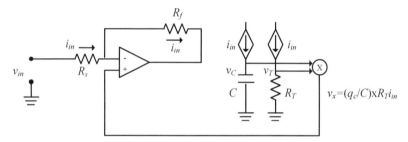

Fig. 1.10. Basic configuration of Kim's memristor [9].

1.4 Structure of the Book

This chapter gives an introduction of memristor and presents a number of memristor emulating circuits. In Chapter 2, we will introduce a memristor based chaotic circuit and attempt to control the chaotic behavior in the circuit using a twin-T notch filter. In Chapter 3, a hyperchaotic memristor based modified canonical Chua's circuit is designed. Chapter 4 presents a design of memristor emulator using light dependent resistor (LDR). Chapters 5 and 6 present memcapacitor emulators using the concept of transformation of memristor. Chapter 7 reports some nonlinear phenomena observed in memristor coupled oscillators. Lastly, Chapter 8 gives a conclusion and suggestions for future work.

References

1. L.O. Chua, "Memristor- The missing element," *IEEE Transactions on Circuit Theory* **18**, 5, pp. 507–513, (1971).
2. D.B. Strukov, G.S. Snider, G.R. Stewart and R.S. Williams, "The missing memristor found," *Nature* **453**, 7191, pp. 80–83, (2008).
3. L.O. Chua and S.M. Kang, "Memristive devices and systems," *Proceedings of the IEEE* **64**, 2, pp. 209–223, (1976).
4. J. Valsa, D. Biolek and Z. Biolek, "An analog model of the memristor," *International Journal of Numerical Modeling: Electronic Networks, Devices and Fields* DOI:10.1002/jnm.786. (2010).
5. Y.V. Pershin and M. Di Ventra, "Practical approach to programmable analog circuits with memristors," *IEEE Transactions on Circuits and Systems I: Regular Papers* **57**, 8, pp. 1857–1864, (2010).
6. J.J. Yang, M.D. Pickett, X. Li, D.A.A. Ohlberg, D.R. Stewart and R.S. Williams, "Memristive switching mechanism for metal/oxide/metal nanodevices," *Nat. Nanotechnol.* **3**, 7, pp. 429–433, (2008).
7. S. Liu, N. Wu and A. Ignatiev, "Electric-pulse-induced reversible resistance change effect in magnetoresistive films," *Appl. Phys. Lett.* **76**, 19, pp. 2749–2751, (2000).
8. B. Muthuswamy and L.O. Chua, "Simplest chaotic circuit," *International Journal of Bifurcation and Chaos* **20**, 5, pp. 1567–1580, (2010).
9. H. Kim, M. Pd. Sah, C. Yang, S. Cho and L.O. Chua, "Memristor emulator for memristor circuit applications," *IEEE Transactions on Circuits and Systems I: Regular Papers* **59**, 11, (2012).
10. M. Pd. Sah, C. Yang, H. Kim and L.O. Chua, "A voltage mode memristor bridge synaptic circuit with memristor emulators," *Sensors*, 12, pp. 3587–3604, (2012).
11. H. Kim, M. Pd. Sah, C. Yang, T. Roska and L.O. Chua, "Neural synaptic weighting with a pulse-based memrsitor circuit," *IEEE Transactions on Circuits and Systems I: Regular Papers* **58**, 1, pp. 148–158, (2011).

Chapter 2

Controlling Chaos in a Memristor Based Circuit

After the successful solid state implementation of memristors, a lot of attention has been drawn to the study of memristor based chaotic circuits. In this chapter, an example of memristor based chaotic circuit is shown. A systematic study of chaotic behavior in such system is performed with the help of nonlinear tools such as bifurcation diagrams, power spectrum analysis and Lyapunov exponents. In particular, a Twin-T notch filter feedback controller is designed and employed to control the chaotic behavior in the memristor based chaotic circuit.

2.1 Introduction

Memristors have been well known as the missing fourth element postulated by L.O. Chua in 1971 [1] until researchers in Hewlett-Packard announced a solid state implementation of memristors in 2008 [2]. Memristor is a two terminal element with variable resistance called memristance which depends on how much electric charge has been passed through it in a particular direction. In other words, memristors have the distinctive ability to memorize the past quantity of electric charge. Therefore, appreciable research interests have been inspired by the successful fabrication of memristors because of the potential applications in computers, neural networks, analog circuitries and so on [3–7].

In [8–9], a four dimensional memristor based chaotic circuit (MCC) is presented by replacing the nonlinear element in Chua's circuit with a memristor, and rich nonlinear dynamical behavior has been observed in the system. In their work, a piece-wise linear function is used to emulate the relation between magnetic flux and electric charge. In [10], a circuit which possesses cubic nonlinear characteristics is employed to implement the MCC. Also, the simplest memristor based chaotic circuit was recently presented in [11]. From the perspective of nonlinear dynamics, controlling chaos in a MCC is attractive and challenging. However, up to now, there are very few research reports addressing chaos control issues in MCC. Applications of chaos control have been proposed in diverse areas of research such as biology, laser physics, chemical engineering, physiology, electric power systems, fluid mechanics, electronics, communications and so on. Chaos control has become a process that manages the dynamics of a nonlinear system on a wider scale to achieve desired goals. The recently developed memristor based circuit is a very good candidate for periodic or chaotic signal generation. The study of chaos control in the MCC provides the designer with an exciting variety of properties, richness of flexibility and opportunities.

There is an expectation that memristors will find use in numerous applications, once their properties have been fully explored and exploited. A number of these applications will be nonlinear and the possibility of experiencing instabilities and chaotic behavior must be considered in these cases. It is important to develop effective methods to control these behaviors concurrently with the development of applications involving memristors rather than as an afterthought or remedy to compensate for some undesired or unexpected performance from a circuit.

This chapter discusses a notch filter feedback controller for a MCC and is organized as follows. Section 2.2 describes the dynamical behavior in MCC using phase portraits, power spectrum diagrams, bifurcation diagrams and Lyapunov exponents. The design of a practical notch filter controller is addressed in Sec. 2.3. Section 2.4 shows the validity of the notch filter feedback control method based on simulation. In Sec. 2.5, experimental measurements from a laboratory prototype are used to validate the theoretical results and numerical simulations. The conclusion is given in Sec. 2.6.

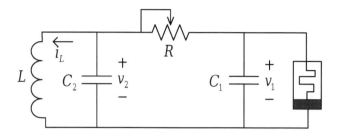

Fig. 2.1. Memristor based chaotic circuit.

2.2 Memristor Based Chaotic Circuit

In this section, a quick review is presented of the dynamics of the MCC proposed in [8–10]. Replacing the Chua diode with a flux-controlled memristor, as shown in Fig. 2.1, the equations for MCC can be written as

$$
\begin{cases}
\dfrac{d\phi(t)}{dt} = -\tau v_1(t) \\[2mm]
\dfrac{dv_1(t)}{dt} = \dfrac{1}{C_1}\left(\dfrac{v_2(t)-v_1(t)}{R}\right) - W(\phi(t))v_1(t) \\[2mm]
\dfrac{dv_2(t)}{dt} = \dfrac{1}{C_2}\left(\dfrac{v_1(t)-v_2(t)}{R}\right) - i_L(t) \\[2mm]
\dfrac{di_L(t)}{dt} = \dfrac{v_2(t)}{L}
\end{cases}
\tag{2.1}
$$

where $\phi(t)$ denotes the magnetic flux between two terminals of a memristor. $\phi(t)$ is the integral of $v_1(t)$ with integral time constant $-\tau$. $W(\phi)$ is the memductance function given by [10]

$$
W(\phi(t)) = a + 3b\phi(t)^2.
\tag{2.2}
$$

The equilibrium points of Eq. (2.1) can be found by setting the left-hand side to zero. Obviously,

$$
P_0 = (^0\phi, {}^0v_1, {}^0v_2, {}^0i_L) = (\Phi_0, 0, 0, 0)
\tag{2.3}
$$

is one equilibrium of the system, where Φ_0 is uncertain but constant. The Jacobian matrix of system (2.1) can be described as

$$J = \begin{bmatrix} 0 & -\tau & 0 & 0 \\ -\dfrac{6b\phi(t)v_1(t)}{C_1} & \dfrac{-1-R(a+3b\phi(t)^2)}{RC_1} & \dfrac{1}{RC_1} & 0 \\ 0 & \dfrac{1}{RC_2} & -\dfrac{1}{RC_2} & -\dfrac{1}{C_2} \\ 0 & 0 & \dfrac{1}{L} & 0 \end{bmatrix}. \qquad (2.4)$$

At equilibrium point P_0, the Jacobian matrix becomes

$$J(P_0) = \begin{bmatrix} 0 & -\tau & 0 & 0 \\ 0 & \dfrac{-1-R(a+3b\Phi_0^2)}{RC_1} & \dfrac{1}{RC_1} & 0 \\ 0 & \dfrac{1}{RC_2} & -\dfrac{1}{RC_2} & -\dfrac{1}{C_2} \\ 0 & 0 & \dfrac{1}{L} & 0 \end{bmatrix}. \qquad (2.5)$$

The system characteristic equation can be written as

$$\lambda(\lambda^3 + \alpha_2\lambda^2 + \alpha_1\lambda + \alpha_0) = 0 \qquad (2.6)$$

where

$$\alpha_2 = \frac{1}{RC_2} + \frac{1+R(a+3b\Phi_0^2)}{RC_1}, \qquad (2.7)$$

$$\alpha_1 = \frac{1}{LC_2} + \frac{(a+3b\Phi_0^2)}{RC_1C_2}, \qquad (2.8)$$

$$\alpha_0 = \frac{1+R(a+3b\Phi_0^2)}{RLC_1C_2}. \qquad (2.9)$$

Table 2.1. Simulation parameters.

Simulation parameters	Values
Inductance L	10.8 mH
Capacitance C_1, C_2	6.8 nF, 268 nF
Resistance R	1800 Ω
Time constant τ	1220
a	0.0007
b	0.00004

P_0 is defined as an unstable equilibrium when at least one root of Eq. (2.6) has the real part greater than zero. Apparently, the coefficients of the cubic polynomial inside the brackets of Eq. (2.6) are all non-zero. Taking advantage of the values of system parameters shown in Table 2.1 and Routh-Hurwitz condition, one or more roots of this cubic polynomial equation have positive real parts when

$$|\Phi_0| < 1.097 \text{ or } 1.154 < |\Phi_0| < 2.364 . \tag{2.10}$$

Equation (2.10) demonstrates that the dynamical behavior of this MCC system is decided by the initial values of state variable $\phi(t)$ [12]. System (2.1) is unstable when the value of Φ_0 satisfies condition (2.10). For the sake of simplicity, only a special condition $\Phi_0 = 0$ is taken into account here. The four characteristic multipliers of Eq. (2.5) are

$$\lambda_1 = 0, \ \lambda_2 = 32462.22 ,$$

$$\lambda_{3,4} = -2458.06 \pm j17521.03 . \tag{2.11}$$

From the values of characteristic multipliers in phase plane, it can be clearly seen that the equilibrium point P_0 is not stable. Due to the existence of the zero characteristic multiplier, it is worth noting that there is no stable equilibrium point in system (2.1). Here, we simulate the system with different values of R. The initial values of state variables are $\phi(0) = 0$ Wb, $v_1(0) = 0.1$ V, $v_2(0) = 0$ V, $i_L(0) = 0$ A. Figure 2.2

shows the strange attractors of MCC when $R = 1800\ \Omega$. Calculation of the Lyapunov exponent is widely used to indicate the existence of chaos [13]. If one Lyapunov exponent is greater than zero and sum of all Lyapunov exponents is less than zero, the system is known to exhibit chaos. We make use of Wolf's method to calculate Lyapunov exponents [13–14]. The results are (0, 3246.54, −2457.24, −2458.05), from which we can see clearly the system is in chaotic state. When the value of resistor R is tuned to $1600\ \Omega$, periodic orbit can be observed as shown in Fig. 2.3.

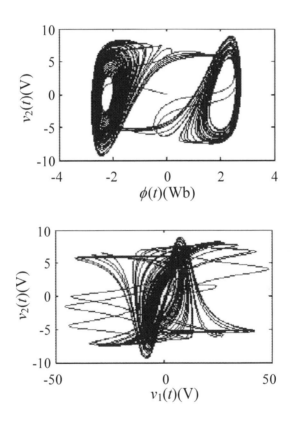

Fig. 2.2. Phase portraits of chaotic attractor, $R = 1800\ \Omega$.

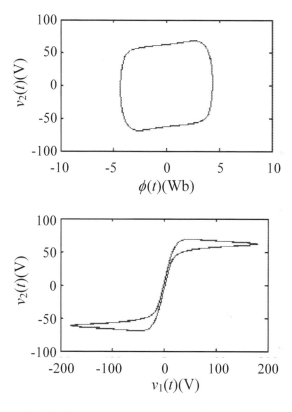

Fig. 2.3. Phase portraits of period-1 orbit, $R = 1600\ \Omega$.

Power spectrum analysis is another simple and fast way to reveal the dynamical behavior of the system [15–17]. Power spectrum distribution can be used to demonstrate the difference between chaotic and periodic signal. Figures 2.4 (a)–(b) show the power spectrum diagrams at $R = 1800\ \Omega$ and $R = 1600\ \Omega$ respectively. From Fig. 2.4 (a), we can see that when the system operates at chaotic state, the power spectrum diagram has complicated broadband and continuous background and the power is mainly gathered in a frequency range from 0 to 55 kHz. In Fig. 2.4 (b), the power spectrum of periodic state is characterized with sharp peaks at certain frequency and its harmonics. As shown in Fig. 2.4 (b), the first sharp peak is located at about 2.88 kHz.

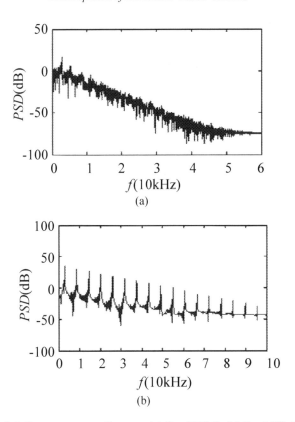

(a)

(b)

Fig. 2.4. Power spectrum diagrams (a) $R = 1800\ \Omega$, (b) $R = 1600\ \Omega$.

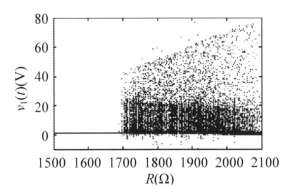

Fig. 2.5. Bifurcation diagrams as R varies.

In order to quickly check the dynamical performances when resistance R changes, we increase this parameter gradually and the steady state of the sampled data are then collected at $v_2(t) = 0$ instants. With sufficient quantity of sampled data, we can construct the bifurcation diagrams of $\phi(t)$ and $v_1(t)$ as depicted in Fig. 2.5, which shows clearly that the output voltage settles at period-1 steady state at first and then jumps to chaotic oscillation abruptly.

2.3 Notch Filter Feedback Controller

The notch filter can remove a "single" frequency from a signal [18–19] and has been employed to control chaos in different nonlinear systems. Notch filter based control is chosen in the study because it is an efficient method for stabilizing steady states (either periodic orbit or fixed point) of chaotic systems. It has the advantages of being easily implemented using analog and digital hardware, and its ability to suppress a selected sub-harmonic component of the input signal. However, previous works are mostly restricted to simulation study [20–23]. In this section, we propose a practical Twin-T notch filter feedback method to control chaos in MCC. The design can be divided into two steps: first we build a Twin-T notch filter with adjustable quality factor, and then we present the design of interface circuit for the filter and MCC system.

2.3.1 *Twin-T notch filter*

Figure 2.6 shows a simple Twin-T notch filter. The state equations for Twin-T notch filter in Fig. 2.6 can be given a

$$
\begin{cases}
\dfrac{dv_{n1}}{dt} = q\,\dfrac{dv_{in}}{dt} + \dfrac{0.5v_{in} + (2q^2 - 2q + 0.5)v_o + (2q - 1)v_{n1} - 2qv_{n2}}{R_n C_n} \\[3mm]
\dfrac{dv_{n2}}{dt} = \dfrac{dv_{in}}{dt} - \dfrac{2v_{n2} - (2q - 1)v_o - v_{n1}}{R_n C_n} \\[3mm]
\dfrac{dv_o}{dt} = \dfrac{dv_{in}}{dt} - \dfrac{2v_{n2} - 2(q - 1)v_o - 2v_{n1}}{R_n C_n}
\end{cases}
$$

$$(2.12)$$

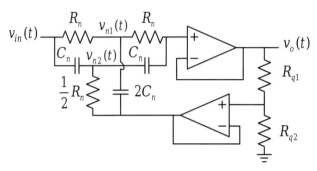

Fig. 2.6. Twin-T notch filter.

where q is quality factor and can be calculated as

$$q = \frac{R_{q2}}{R_{q1} + R_{q2}}. \tag{2.13}$$

$v_{in}(t)$ and $v_o(t)$ are the instantaneous input and output signal respectively, $v_{n1}(t)$ and $v_{n2}(t)$ are the internal state variables in the filter circuit. Making use of Laplace transform, we can get the input-output transfer function of (2.12).

$$F(s) = \frac{V_o(s)}{V_{in}(s)} = \frac{s^2 + \dfrac{1}{R_n^2 C_n^2}}{s^2 + \dfrac{4}{R_n C_n}(1 - q)s + \dfrac{1}{R_n^2 C_n^2}}. \tag{2.14}$$

By letting Eq. (2.14) equal to zero, the notch frequency f can be calculated as

$$Ff = \frac{1}{2\pi R_n C_n}. \tag{2.15}$$

Equation (2.14) also shows that the width of the notch is determined by q. If $q = 1$, the notch becomes so narrow that $F(s) = 1$. Likewise, the notch becomes wider as q decreases. Letting $R_n = 10$ k$\square$, $C_n = 10$ nF, $q = 0.8$, the Bode plot of Eq. (2.14) is given in Fig. 2.7 which gives the

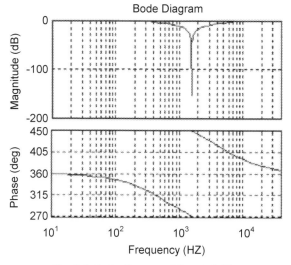

Fig. 2.7. Bode diagram of the notch filter.

magnitude and phase response characteristics of the designed filter. As the name suggests, a "notch" is observed at the notch frequency in the Bode plot. The notch filter can greatly decay the sub-harmonic component of its input signal at notch frequency.

2.3.2 *Interface circuit*

The input and output of the notch filter discussed above are both voltage signals. In practice, the output signal cannot be easily fed back to MCC system directly. It is necessary to exploit a linear voltage to current converter that is capable of sinking a current control signal into one node of MCC without changing the characteristics of the output of notch filter. Figure 2.8 shows a simple voltage to current converter only requiring one amplifier and five resistors [24].

Assuming the output current and interface voltage of this converter are $i_o(t)$ and $v_o(t)$, respectively. Based on the basic principle of ideal amplifier, we can get the following relation

$$i_o(t) + \frac{R_2 R_4 - R_1(R_3 + R_5)}{R_1 R_3 (R_4 + R_5)} v_c(t) = -\frac{R_2}{R_1 R_3} v_o(t). \qquad (2.16)$$

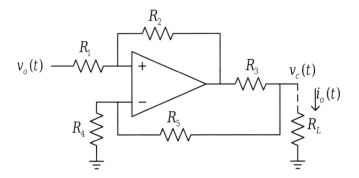

Fig. 2.8. Voltage to current converter.

Equation (2.16) indicates that $i_o(t)$ depends on not only input voltage $v_o(t)$ but also on the interface voltage and five resistors in a complicated relation and the value of load resistance. This converter is designed to excite the MCC system, which means the load should not affect the voltage to current characteristic. Therefore, the voltage to current characteristic drawn by Eq. (2.16) is not a linear conversion unless the following equation holds

$$R_2 R_4 - R_1 (R_3 + R_5) = 0. \tag{2.17}$$

Based on the condition (2.17), we can get the input and output relation for the converter

$$i_o(t) = -\frac{R_2}{R_1 R_3} v_o(t). \tag{2.18}$$

In particular, by letting $R_5 = 0$, $R_1 = R_2$ and $R_4 = R_3$ a simpler relation can be obtained

$$i_o(t) = -\frac{1}{R_3} v_o(t). \tag{2.19}$$

R_3 determines the effective strength of feedback operation. It should be noted that the maximum output current of the converter is limited by the saturated output voltage of the amplifier. In other words, Eq. (2.19) does not hold when the output voltage of the amplifier is in saturation.

2.4 Simulation Results

In this section, we connect the notch filter controller to the MCC system and present the scenario of chaos control in simulation. The control strategy of the notch filter feedback controller is that the feedback will cause all oscillations to decay except for that corresponding to the notch frequency which is not affected by the feedback process. Thus, the notch filter will result in a stabilized periodic solution [21]. In particular, we take the state variable $v_2(t)$ as the input of notch filter controller. Then, $v_2(t)$ goes through the Twin-T notch filter and voltage-to-current converter, and $i_o(t)$ is fed back to the junction of R and C_1. Figure 2.9 gives the full schematic of the entire system.

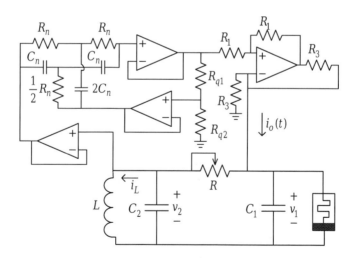

Fig. 2.9. Schematic of the MCC with notch filter controller.

The state equations of the new system can be organized as

$$
\begin{cases}
\dfrac{dv_{n1}(t)}{dt} = \dfrac{q}{C_2}(\dfrac{v_1(t)-v_2(t)}{R}-i_L(t)) + \dfrac{0.5v_2(t)+(2q^2-2q+0.5)v_o(t)+(2q-1)v_{n1}(t)-2qv_{n2}(t)}{R_nC_n} \\[3mm]
\dfrac{dv_{n2}(t)}{dt} = \dfrac{1}{C_2}(\dfrac{v_1(t)-v_2(t)}{R}-i_L(t)) - \dfrac{2v_{n2}(t)-(2q-1)v_o(t)-v_{n1}(t)}{R_nC_n} \\[3mm]
\dfrac{dv_o(t)}{dt} = \dfrac{1}{C_2}(\dfrac{v_1(t)-v_2(t)}{R}-i_L(t)) - \dfrac{2v_{n2}(t)-2(q-1)v_o(t)-2v_{n1}(t)}{R_nC_n} \\[3mm]
\dfrac{d\phi(t)}{dt} = -\tau v_1(t) \\[3mm]
\dfrac{dv_1(t)}{dt} = \dfrac{1}{C_1}(\dfrac{v_2(t)-v_1(t)}{R}) - \dfrac{1}{R_3}v_o(t)-(\alpha+3\beta\phi(t)^2)v_1(t)) \\[3mm]
\dfrac{dv_2(t)}{dt} = \dfrac{1}{C_2}(\dfrac{v_1(t)-v_2(t)}{R}-i_L(t)) \\[3mm]
\dfrac{di_L(t)}{dt} = \dfrac{v_2(t)}{L}
\end{cases}
$$

$$(2.20)$$

The state of system (2.20) can be controlled by tuning the parameters in notch filter, such as quality factor q, notch frequency f or control coefficient R_3. The Jacobian matrix of (2.20) at equilibrium point

$$P_{c0} = ({}^0v_{n1}, {}^0v_{n2}, {}^0v_o, {}^0\phi, {}^0v_1, {}^0v_2, {}^0i_L) = (0,0,0,0,0,0,0) \qquad (2.21)$$

is shown in Eq. (2.23). Note that a zero characteristic multiplier always exists at the equilibrium point, which means the chaos control method used in this paper cannot control the MCC to a stable equilibrium point.

In order to make the notch filter controller effective, the notch frequency is supposed to be less than the maximum possible frequency f_{max} in chaotic signal. Therefore, the value of resistor R_n must satisfy

$$R_n > \frac{1}{2\pi C_n f_{max}}. \qquad (2.22)$$

For the chaotic circuit discussed above, the frequency of the output chaotic signal scatters from 0 to 55 kHz. Assuming the capacitance C_n is fixed on 2.2 nF, the value of R_n can be tuned in the range $(1315, +\infty)\ \Omega$.

Note that when Eq. (2.22) is not held, the control method proposed in this paper is equivalent to the proportional feedback control because the notch filter does not act on the feedback signal.

$$J_c(P_{c0}) = \begin{bmatrix} \dfrac{2q-1}{R_nC_n} & \dfrac{-2q}{R_nC_n} & \dfrac{2q^2-2q+0.5}{R_nC_n} & 0 & \dfrac{q}{RC_2} & \dfrac{-q}{RC_2}+\dfrac{0.5}{R_nC_n} & -\dfrac{q}{C_2} \\[2ex] \dfrac{1}{R_nC_n} & \dfrac{-2}{R_nC_n} & \dfrac{2q-1}{R_nC_n} & 0 & \dfrac{1}{RC_2} & \dfrac{-1}{RC_2} & -\dfrac{1}{C_2} \\[2ex] \dfrac{2}{R_nC_n} & \dfrac{-2}{R_nC_n} & \dfrac{2(q-1)}{R_nC_n} & 0 & \dfrac{1}{RC_2} & \dfrac{-1}{RC_2} & \dfrac{1}{C_2} \\[2ex] 0 & 0 & 0 & 0 & -\dfrac{1}{R_6C_3} & 0 & 0 \\[2ex] 0 & 0 & -\dfrac{1}{R_3C_1} & 0 & \dfrac{-1-R\alpha}{RC_1} & \dfrac{1}{RC_1} & 0 \\[2ex] 0 & 0 & 0 & 0 & \dfrac{1}{RC_2} & -\dfrac{1}{RC_2} & -\dfrac{1}{C_2} \\[2ex] 0 & 0 & 0 & 0 & 0 & \dfrac{1}{L} & 0 \end{bmatrix}.$$

$$(2.23)$$

Letting R_n = 25 kΩ, q = 0.9, R_3 = 4 kΩ, according to Eq. (2.15), the notch frequency is 2.9 kHz approximately. Then, we connect the controller at time t = 12 ms and the results are shown in Fig. 2.10, in which red step line represents the control signal. The simulation results show that after a short period of transient, the chaotic signal in MCC can be stabilized at period-1 state, due to the notch filter controller. Furthermore, substituting all parameter values into Eq. (2.23), the seven characteristic multipliers are

$$\lambda_1 = 0, \lambda_2 = 23789.88, \lambda_3 = -18181.82,$$

$$\lambda_{4,5} = -3865.33 \pm j19397.54,$$

$$\lambda_{6,7} = -2081.54 \pm j15719.09.$$

$$(2.24)$$

The second characteristic multiplier of Eq. (2.24) is on the right half phase plane, which reveals that the equilibrium point P_{c0} after control is not stable.

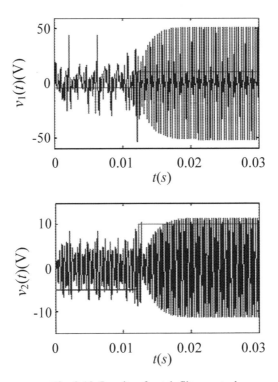

Fig. 2.10. Results of notch filter control.

2.5 Experimental Setup

Based on the schematic in Fig. 2.9, we have built a MCC system with a notch filter controller as shown in Fig. 2.11. This experimental prototype is comprised of three parts: MCC, notch filter and voltage-to-current converter. The MCC system is built referring to the schematic in [10]. The parameters of experimental components are given in Table 2.2.

Table 2.2. Experimental parameters.

Experimental parameters	Values
Inductance L	10.8 mH
Capacitor C_1, C_2	6.8 nF, 268 nF
Time constant τ	1220
a	0.00067
b	0.0000044
Capacitor C_n	2.2 nF
Resistor R_n	25 k□
Resistor R	(0~4000) □
Resistor R_3	6.6 k□
Resistor R_1	100 k□
q	0.92

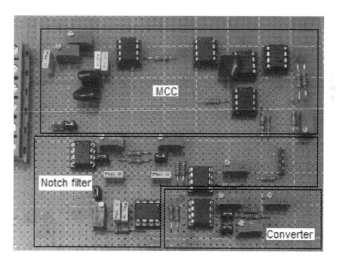

Fig. 2.11. Experimental prototype.

First, the notch filter controller is disconnected from the MCC, period-1 orbit is captured and exhibited in Figs. 2.12 (a) and (b) at $R = 1500$ □. Meanwhile, chaotic state is shown in Figs. 2.12 (c) and (d) when $R = 1890$ □. The experimental results reveal that the system exhibits periodic and chaotic behavior. It should be emphasized that only period-1 and chaotic states are found in experimental prototype by tuning resistor R, as discussed in Sec. 2.2 by simulation.

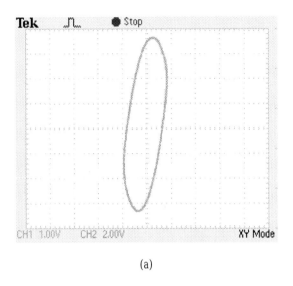

(a)

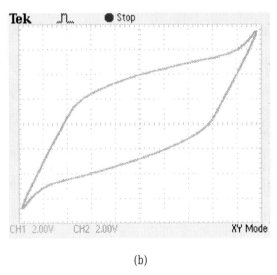

(b)

Fig. 2.12. Phase portraits. (a) and (b) are period-1 orbit, while (c) and (d) show chaotic attractor. The vertical axis for all the above diagrams is $v_2(t)$. For (a) and (c) the horizontal axis is $\phi(t)$. The horizontal axis of (b) and (d) is $v_1(t)$.

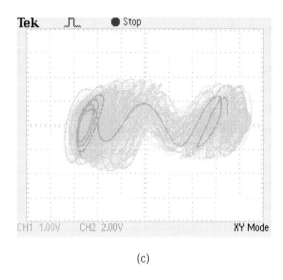

(c)

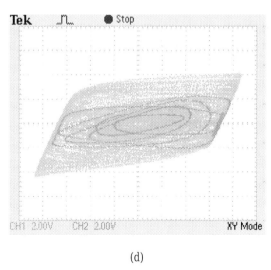

(d)

Next, we test the effectiveness of the designed notch filter controller. The moment of notch filter controller connection is denoted by a step signal as shown in Fig. 2.13. From Fig. 2.13 (a) we can see clearly that after about 40 ms transient, $v_1(t)$ is transformed from chaotic into periodic signal. Figure 2.13 (b) shows the transient process of $v_2(t)$. Apart

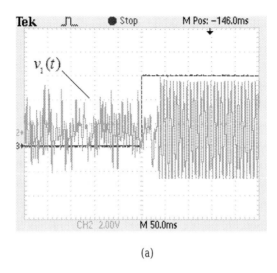

(a)

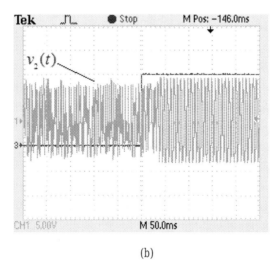

(b)

Fig. 2.13. Time-domain waveforms showing the effectiveness of the notch filter controller.

from the time-domain waveforms, we also capture the experimental power spectrum diagram. The power spectrum diagrams before and after the controller is connected can be obtained by FFT function in oscilloscope TDS20114. In particular, the power spectrum of state variable $v_1(t)$ is given in Fig. 2.14. The continuous broadband power spectrum in Fig. 2.14 (a) indicates the existence of chaotic state. Its maximum peak value is 10 dB, less than that of periodic spectrum shown in Fig. 2.14 (b). The first sharp peak in Fig. 2.14 (b) can be located at 2.71 kHz. These results are in good agreement with simulation.

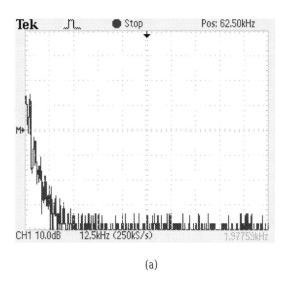

(a)

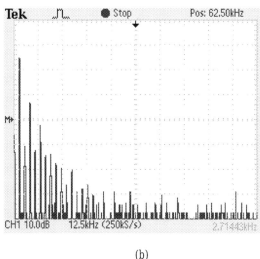

(b)

Fig. 2.14. Power spectrum diagrams (a) before connection, (b) after connection.

2.6 Conclusion

In this chapter, the periodic and chaotic trajectories of the MCC [8–10] are shown and the Lyapunov exponents are calculated to confirm the emergence of chaos. The bifurcation diagrams indicate that MCC system jumps from period-1 to chaos abruptly as resistance R increases. A controller consisting of a Twin-T notch filter and a voltage-to-current converter is designed, and is employed to control chaos in the MCC system. The notch filter and voltage-to-current converter used in this paper are convenient to implement. Simulation and experimental results both manifest that the proposed controller can control the MCC system from chaos into period-1 state under certain control parameters.

References

1. L.O. Chua, "Memristor- The missing element," *IEEE Transactions on Circuit Theory* **18**, 5, pp. 507–513, (1971).
2. D.B. Strukov, G.S. Snider, G.R. Stewart and R.S. Williams, "The missing memristor found," *Nature* **453**, 7191, pp. 80–83, (2008).
3. Y. Ho, G.M. Huang and P. Li, "Nonvolatile memristor memory: device characteristics and design implications," *IEEE/ACM 2009 International Conference on Computer-Aided Design*, pp. 485–490, (2009).
4. F.Z. Wang, N. Helian, S. Wu, M.G. Lim, Y. Guo and M.A. Parker, "Delayed switching in memristors and memristive systems," *IEEE Electron. Device Letters* **31**, 7, pp. 755–757, (2010).
5. Y.V. Pershin and M. Di Ventra, "Experimental demonstration of associative memory with memristive neural networks," (2009) [Online]. Available: http://arXiv.org/abs/arXiv:0905.2935
6. H. Choi, H. Jung, J. Lee, J. Yoon, J. Park, D.-J. Seong, W. Lee, M. Hasan, G.-Y. Jung and H.H. Hwang, "An electrically modifiable synapse array of resistive switching memory," *Nanotechnology* **20**, 345201, (2009).
7. Y.V. Pershin and M. Di Ventra, "Practical approach to programmable analog circuits with memristors," *IEEE Transactions*

on *Circuits and Systems I: Regular Papers* **57**, 8, pp. 1857–1864, (2010).

8. B. Muthuswamy and P.P. Kokate, "Memristor-based chaotic circuits," *IETE Technical Review* **26**, 6, pp. 417–429, (2009).

9. M. Itoh and L.O. Chua, "Memristor oscillators," *International Journal of Bifurcation and Chaos* **18**, 11, pp. 3183–3206, (2008).

10. B. Muthuswamy, "Implementing memristor based chaotic circuits," *International Journal of Bifurcation and Chaos* **20**, 5, pp. 1335–1350, (2010).

11. B. Muthuswamy and L.O. Chua "Simplest chaotic circuit," *International Journal of Bifurcation and Chaos* **20**, 5, pp.1567–1580, (2010).

12. B.C. Bao, Z. Liu and J.P. Xu, "Steady periodic memristor oscillator with transient chaotic behaviours," *Electronics Letters* **46**, 3, pp. 237–238, (2010).

13. A. Wolf, J. Swift, H. Swinney and J. Vastano, "Determining Lyapunov exponents from a time series," *Physica D: Nonlinear Phenomena* **16**, 3, pp. 285–317, (1985).

14. S. Siu, "Lyapunov exponent toolbox," http://www.mathworks.com/matlabcentral/fileexchange/.

15. N.F. Rulkov and A.R. Volkovskii, "Generation of broad-band chaos using blocking oscillator," *IEEE Transactions on Circuits and Systems I: Fundamental Theory and Applications* **48**, 6, pp. 673–679, (2001).

16. T. Matsumoto, L.O. Chua and M. Komuro, "The double scroll," *IEEE Transactions on Circuits and Systems* **32**, 8, pp. 797–818, (1985).

17. H. Li, Z. Li, W.A. Halang, B. Zhang and G. Chen, "Analyzing chaotic spectra of dc–dc converters using the Prony method," *IEEE Transactions on Circuits and Systems II: Express Briefs* **54**, 1, pp. 61–65, (2007).

18. A. Kircay and U. Cam, "Differential type class-AB second-order log-domain notch filter," *IEEE Transactions on Circuits and Systems I: Regular Papers* **55**, 5, pp. 1203–1212, (2008).

19. M. Mojiri and A.R. Bakhshai, "Estimation of n frequencies using adaptive notch filter," *IEEE Transactions on Circuits and Systems II: Express Briefs* **54**, 4, pp. 338–342, (2007).

20. C. Cai, Z. Xu, W. Xu and B. Feng, "Notch filter feedback control in a class of chaotic systems," *Automatica* **38**, 4, pp. 695–701, (2002).

21. A.A. Zaher and A. Abu-Rezq, "Tuning of notch filters for controlling chaos in a Chua's circuit," *IEEE International Conference on Signal Processing and Communications*, pp. 305–308, (2007).

22. A.M. Athalye and W.J. Grantham, "Notch filter feedback control of a chaotic system," *American Control Conference*, pp. 837–841, (1995).

23. A. Ahlborn and U. Parlitz, "Chaos control using notch filter feedback," *Physical Review Letters* **96**, 3, 034102, (2006).

24. C.D. Johnson, *Process Control Instrumentation Technology* (8th Edition), Prentice Hall, pp. 97–98, (2005).

Chapter 3

Hyperchaos in a Memristor Based Modified Canonical Chua's Circuit

In this chapter, a memristor with cubic nonlinear characteristics is employed in the modified canonical Chua's circuit. A systematic study of hyperchaotic behavior in this circuit is performed with the help of nonlinear tools such as Lyapunov exponents, phase portraits and bifurcation diagrams. In particular, an imitative memristor circuit is examined to reveal the construction of hyperchaotic attractors.

3.1 Introduction

In 2008, researchers in Hewlett-Packard announced that a solid state implementation of memristor has been successfully fabricated [1]. Subsequently, appreciable research interests have been inspired due to the potential applications of memristors in computers, neural networks, secure communications and so on [2-10]. In the last couple of years, the use of memristor based circuits to construct chaotic systems has attracted a lot of interest [11-16].

In [11] and [13], a three dimensional memristor based chaotic circuit (MCC) is presented by replacing the nonlinear element in Chua's circuit with a memristor. In their work, a piece-wise linear function is used to emulate the relation between magnetic flux and electric charge. In [14], a circuit which possesses cubic nonlinear characteristics is employed to implement the MCC. Also, the simplest memristor based chaotic circuit was recently presented in [15].

The circuit under discussion has been derived from that presented in [17] being one of the earliest papers to examine chaos in a memristor based circuit. In [13], an inductor is added and chaos is observed from a

37

similar fourth-order circuit. This paper shows that the inclusion of active element $-G$ enables chaos and hyperchaos to be observed from a fourth-order circuit.

There is an expectation that memristors will find use in numerous applications, once their properties have been fully explored and exploited. A number of these applications will be nonlinear and the possibility of experiencing instabilities and chaotic behavior should be considered in these cases. This article makes the first attempt to construct a memristor based hyperchaotic circuit (MHC). This is a novel circuit to examine and knowledge gained may be expected to contribute to further research into memristors and their applications. The development of reliable hyperchaotic circuits will enable the construction of more complex time domain signals, which may result in applications for secure communications and encryption.

Initial applications for memristors include the replacement of transistors in microprocessors to overcome the technological impasse where transistor size cannot be shrunk any further without suffering from thermal and electrical problems. The intention to use memristors in place of transistors is simply a way of 'buying time', before long memristors used in this context will also reach a minimal functional size and a maximum component-per-chip density. A real leap forward in the evolution of computing can occur when memristors are utilized for both their analog and digital properties concurrently. Analog/digital hybrid computers and artificial neural network computers are expected to deliver a variety of currently unavailable benefits; including artificial intelligence and self-learning. Much work is already underway to develop such applications [18–19].

Hyperchaotic systems are being developed for applications in secure communications [20–21]. It is important that memristive hyperchaotic systems be developed for implementation in coming generations of memristor based devices. This would enable a simple internal means of incorporating secure encryption into any communication system.

In this chapter, hyperchaos in a memristor based modified canonical Chua's circuit is investigated and the rest of this paper is organized as follows. Section 3.2 derives the system state equations. Section 3.3

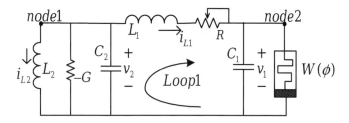

Fig. 3.1. Memristor based modified canonical Chua's circuit.

describes the dynamical behavior in the MHC using a bifurcation diagram, phase portraits and Lyapunov exponents. In Sec. 3.4, experimental measurements from a laboratory prototype are used to validate the theoretical results and numerical simulations. Conclusions are given in Sec. 3.5.

3.2 Dynamical Equations of MHC

By replacing the nonlinear element with a flux-controlled memristor in canonical Chua's circuit [17], a memristor based canonical Chua's circuit can be obtained [11]. This circuit's chaotic attractor has a single positive Lyapunov exponent. Furthermore, we modify this circuit by adding an inductor in parallel with conductance $-G$ and a fourth-order memristor based canonical oscillator is established, as shown in Fig. 3.1.

Due to the presence of the flux-controlled memristor, we define the state variables using charge and flux for the oscillator instead of voltage and current

$$x(t) = \begin{Bmatrix} q_{l2}(t) \\ \phi_{c2}(t) \\ q_{l1}(t) \\ \phi_{c1}(t) \end{Bmatrix} \tag{3.1}$$

where $\phi_{c1}(t)$ and $\phi_{c2}(t)$ denote the magnetic flux between the two terminals of the capacitors C_1 and C_2, $q_{l1}(t)$ and $q_{l2}(t)$ denote the electric charge going through the two terminals of the inductors, respectively. Applying Kirchhoff's voltage law at node 1, we can get

$$L_2 \frac{di_{l_2}(t)}{dt} = v_2(t).$$ (3.2)

According to Faraday's law, by integrating Eq. (3.2) the first state equation can be obtained

$$L_2 \frac{dq_{l_2}(t)}{dt} = \phi_{c_2}(t).$$ (3.3)

By making use of Kirchhoff's voltage law in Loop 1 and integrating, we can get the second state equation

$$L_1 \frac{dq_{l_1}(t)}{dt} = \phi_{c_2}(t) - q_{l_1}(t)R - \phi_{c_1}(t).$$ (3.4)

Similarly, by applying Kirchhoff's current law at nodes 1 and 2 and integrating, two more state equations can be given by

$$C_2 \frac{d\phi_{c_2}(t)}{dt} = -q_{l_2}(t) + G\phi_{c_2}(t) - q_{l_1}(t)$$ (3.5)

and

$$C_1 \frac{d\phi_{c_1}(t)}{dt} = q_{l_1}(t)R - q_M(t)$$ (3.6)

where $q_M(t)$ denotes the electric charge going through the two terminals of the memristor, and

$$\frac{dq_M(t)}{dt} = i_M(t) = W(\phi_{c_1}(t))v_1(t).$$ (3.7)

$W(\phi_{c_1}(t))$ is the memductance function described by $W(\phi_{c_1}(t))$. Referring to [13], cubic nonlinear characteristics are employed to address this flux-controlled memristor,

$$q_M(t) = a\phi_{c_1}(t) + b\phi_{c_1}(t)^3.$$ (3.8)

Correspondingly, memductance function $W(\phi_{c1}(t))$ therefore can be given by

$$W(\phi_{c1}(t)) = \frac{dq_M(t)}{d\phi_{c1}(t)} = a + 3b\,\phi_{c1}(t)^2. \tag{3.9}$$

Based on Eqs. (3.2)–(3.8), the equations for the circuit in Fig. 3.1 can be written as,

$$\begin{cases} \dfrac{dq_{l2}(t)}{dt} = \dfrac{1}{L_2}\phi_{c2}(t) \\[2mm] \dfrac{d\phi_{c2}(t)}{dt} = \dfrac{1}{C_2}(-q_{l2}(t) + G\phi_{c2}(t) - q_{l1}(t)) \\[2mm] \dfrac{dq_{l1}(t)}{dt} = \dfrac{1}{L_1}(\phi_{c2}(t) - q_{l1}(t)R - \phi_{c1}(t)) \\[2mm] \dfrac{d\phi_{c1}(t)}{dt} = \dfrac{1}{C_1}(q_{l1}(t)R - a\phi_{c1}(t) - b\phi_{c1}(t)^3) \end{cases} \tag{3.10}$$

The equilibrium points of Eq. (3.10) can be found by setting the left-hand side to zero. Obviously,

$$P_0 = (\,^0\phi_{c1}, {}^0\phi_{c2}, {}^0q_{l1}, {}^0q_{l2}) = (0,0,0,0) \tag{3.11}$$

is one equilibrium of the system. The Jacobian matrix of system (3.10) at equilibrium P_0 can be described as

$$J(P_0) = \begin{bmatrix} 0 & \dfrac{1}{L_2} & 0 & 0 \\[3mm] -\dfrac{1}{C_2} & \dfrac{G}{C_2} & -\dfrac{1}{C_2} & 0 \\[3mm] 0 & \dfrac{1}{L_1} & -\dfrac{R}{L_1} & -\dfrac{1}{L_1} \\[3mm] 0 & 0 & \dfrac{R}{C_1} & 0 \end{bmatrix}. \tag{3.12}$$

Based on Eq. (3.12), the characteristic polynomial of the linearized system can be written as,

$$\lambda^4 + \alpha_4 \lambda^3 + \alpha_3 \lambda^2 + \alpha_2 \lambda + \alpha_1 = 0 \qquad (3.13)$$

where

$$\alpha_4 = \frac{R}{L_1} - \frac{G}{C_2},$$

$$\alpha_3 = \frac{R}{C_1 L_1} - \frac{RG}{C_2 L_1} + \frac{1}{L_1 C_2} + \frac{1}{L_2 C_2},$$

$$\alpha_2 = \frac{R}{C_2 L_1 L_2} - \frac{GR}{L_1 C_1 C_2},$$

$$\alpha_1 = \frac{1}{L_1 L_2 C_1 C_2}.$$

Using the values of system parameters shown in Table 3.1, the four characteristic multipliers of Eq. (3.13) are

$$\lambda_{1,2} = -3392.642 \pm j1017873.005,$$

$$\lambda_{3,4} = 16559.309 \pm j28805.424. \qquad (3.14)$$

From the values of the characteristic multipliers, we can see clearly that the equilibrium point P_0 is not stable. Therefore, it is possible to produce chaos or hyperchaos in the system described by Eq. (3.10).

3.3 Simulation Results

Due to the inconvenience of experimentally measuring electric charge accurately, voltage and current are employed again as state variables in this section in order to provide a comprehensive investigation and analysis using simulation and experimental setup. Hence, system Eq. (3.10) can be reorganized by

$$\begin{cases} \dfrac{d\phi_{c1}(t)}{dt} = -\tau v_1(t) \\[2mm] \dfrac{dv_1(t)}{dt} = \dfrac{1}{C_1}(i_{L1} - W(\phi_{c1}(t))v_1(t)) \\[2mm] \dfrac{dv_2(t)}{dt} = \dfrac{1}{C_2}(Gv_2(t) - i_{L1}(t) - i_{L2}(t)) \\[2mm] \dfrac{di_{L1}(t)}{dt} = \dfrac{1}{L_1}(v_2(t) - v_1(t) - Ri_{L1}(t)) \\[2mm] \dfrac{di_{L2}(t)}{dt} = \dfrac{1}{L_2}v_2(t) \end{cases} \qquad (3.15)$$

where an integration constant τ is introduced to rescale the values of voltage into practical range. We simulate the system on the basis of Eq. (3.15) with values in Table 3.1 and obtain the phase portraits in Fig. 3.2 correspondingly; they indicate the existence of chaos or hyperchaos. For the sake of comparison, two more phase portraits of period-1 behavior are given in Fig. 3.3 at $R = 140$ □.

In order to quickly check the dynamic performances when resistance R changes in the interval $R \in (20,150)$, we increase this parameter gradually and the steady state of the sampled data is then collected at the instants when

$$\phi_{c1}(t) = 0, \quad \frac{d\phi_{c1}(t)}{dt} < 0 . \qquad (3.16)$$

Table 3.1. Simulation parameters.

Simulation parameters	Values
Inductance L_1, L_2	10 mH, 60 mH
Capacitance C_1, C_2	6.8 nF, 15 nF
Resistance R	70 □
Time constant τ	26000
Conductance G	0.0005 S
A	0.00067
B	0.000029

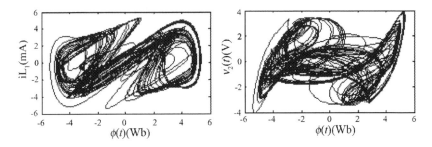

Fig. 3.2. Phase portrait of two-dimensional plane when $R = 70$ □.

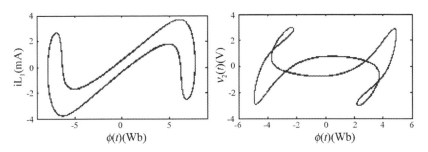

Fig. 3.3. Phase portrait of two-dimensional plane when $R = 140$ □.

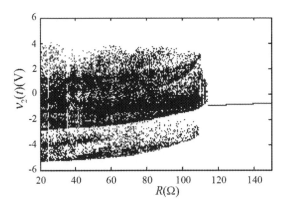

Fig. 3.4. Bifurcation diagram with R as varying parameter.

With a sufficient quantity of sampled data, we can construct the bifurcation diagram of $v_2(t)$ as depicted in Fig. 3.4, which clearly shows that the output voltage settles at oscillation at first and then jumps to period-1 steady state behavior abruptly at $R = 114$ □.

Calculations of Lyapunov exponents are widely used to indicate the existence of chaos or hyperchaos [22–23]. For the fourth-order system of (3.10), assume the four Lyapunov exponents are $Le_m (m = 1, ..., 4)$ and satisfy

$$Le_1 \geq Le_2 > Le_3 > Le_4,$$

$$Le_1 + Le_2 + Le_3 + Le_4 < 0. \tag{3.17}$$

The dynamical behavior in system (3.10) can be classified with respect to Le_m,

- If $Le_1 = 0$, $0 > Le_2 > Le_3 > Le_4$, periodic behavior happens;
- If $Le_1 > 0$, $Le_2 = 0$, $0 > Le_3 > Le_4$, chaotic oscillation happens;
- If $Le_1 > 0$, $Le_2 > 0$, $Le_3 = 0$, $0 > Le_4$, hyperchaotic oscillation happens.

We make use of the singular value decomposition method to calculate Lyapunov exponents [24–25]. The iterative operations are 4000 steps with integration time step 0.0005 s. The numerical calculated result of the three maximum Lyapunov exponents is shown in Fig. 3.5, from which we can clearly see that there are three different dynamic states existing in system (3.10) in accordance to the value of element R. Hyperchaotic oscillation is revealed at $R = 20$ □ due to two positive Lyapunov exponents. The dynamical behavior of system (3.10) evolves from hyperchaos to chaos at $R = 42.5$ □ and back to hyperchaos at $R = 46$ □ as the value of element R is increased. After an interval of hyperchaos, system (3.10) eventually steps into periodic behavior with the maximum Lyapunov exponents very close to 0. Figure 3.5 is in good agreement with the bifurcation diagram. It needs to be stressed that even in the case of periodic behavior; the maximum Lyapunov exponent produced by singular value decomposition method is not accurately equivalent to 0 due to the computing error caused by numerical computation.

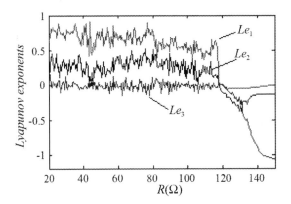

Fig. 3.5. The three maximum Lyapunov exponents.

Table 3.2. Lyapunov exponents and dynamical property of (10).

R	Lyapunov exponents $Le_1{\sim}Le_4$	Property
20	0.6166, 0.3494, 0.0423, −2.4785	hyperchaos
43	0.5025, 0.0464, −0.0232, −2.9646	chaos
65	0.6111, 0.2684, −0.0492, −2.5851	hyperchaos
80	0.7429, 0.5048, −0.0412, −2.1688	hyperchaos
102	0.5347, 0.2776, −0.0124, −2.2312	hyperchaos
120	0.0056, −0.0115, −0.1121, −2.668	periodicity
145	−0.017, −0.1488, −1.0351, −2.8357	periodicity

The four Lyapunov exponents and dynamical property of (3.10) at some typical values of R are listed in Table 3.2.

3.4 Experimental Setup

Based on the schematic in Fig. 3.1, we have built a MHC system with experimental parameters of the two inductors being: L_1 = 10.7 mH, L_2 = 68.7 mH, the other component parameters are given in Table 3.1. The full schematic of the experimental circuit is shown in Fig. 3.6 and an image of the physical circuit is shown in Fig. 3.7.

In the circuit, a set of inductors are connected in series to obtain the required inductance L_2. The use of inductors in series is known to

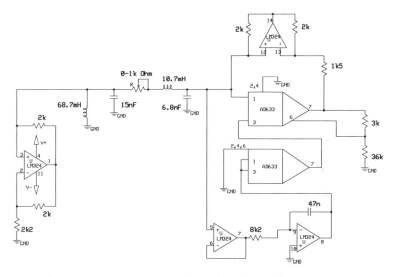

Fig. 3.6. Schematic of the circuit used in the experimental setup.

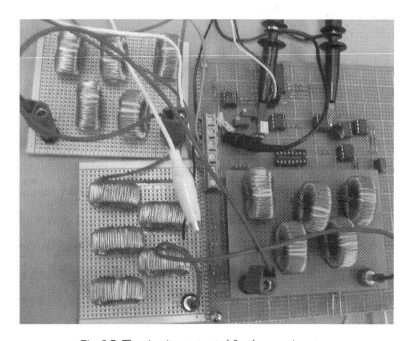

Fig. 3.7. The circuit constructed for the experiment.

introduce parasitic resistive effects into circuits, the components used in the experiment were chosen for their high frequency range and low DC resistance of 0.075 □ to minimize these effects. The coils were connected in a staggered pattern, as shown in Fig. 3.7 to reduce the occurrence of magnetic coupling.

An integration circuit is established to achieve the transformation from voltage $v_1(t)$ into flux $\phi_{c1}(t)$ with integration constant $\tau = 1 / R_{in}C_{in} = 25947$ (see Fig. 3.6). The memristor with cubic nonlinearity in this MHC system is built referring to the schematic studied in [14], using a LM324 quad operational amplifier package, AD633 multipliers and powered by a +/–15V supply. AD633 multipliers produce an output of xy/10 for inputs x & y. When a resistor divider feedback network is added, the gain of the output can be adjusted, in this circuit the output is 3xy/2.

The phase portraits were obtained using an Iwatsu SS-5702 analog oscilloscope. Firstly, by tuning resistor R, two strange attractors are captured and exhibited in Figs. 3.8 (a) and (b) when R = 70 □. It should be emphasized that it is impossible to classify these two strange attractors into chaos or hyperchaos by visual observation.

An effective way to make a distinction between chaotic and hyperchaotic attractor from a prototype of nonlinear circuit is taking advantages of Wolf's method to calculate Lyapunov exponents based on sampled time series [12, 23]. Data acquisition card NI PCI-6284 is introduced to sample output voltage v_2. By making use of time series v_2 (5000 sampling points) and the MATLAB program provided by Govorukhin [26], two positive maximum Lyapunov exponents are obtained, which presents conclusive evidence that the strange attractors given in Figs. 3.8 (a) and (b) can be categorized into hyperchaos. Subsequently, period-1 orbit is exposed in Figs. 3.8 (c) and (d) when R = 140 □. The experimental results reveal that the system has rich dynamical behavior, and are in good agreement with the simulation results obtained in Sec. 3.3. It should be emphasized that only period-1, chaotic and hyperchaotic states are found in experimental prototype by tuning resistor R, as discussed in Sec. 3.3 by simulation.

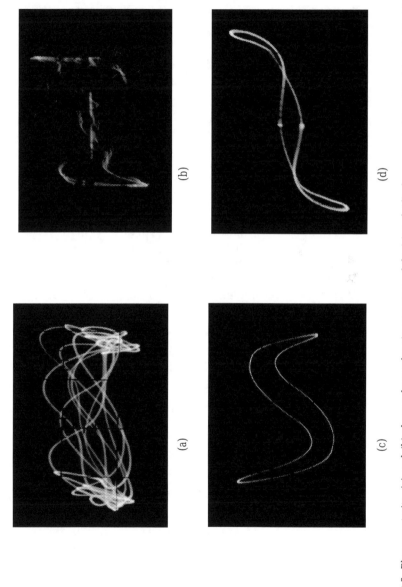

Fig. 3.8. Phase portraits. (a) and (b) show a hyperchaotic attractor, while (c) and (d) show a period-1 orbit. (a) and (c) phi versus iL1 (x-axis: 1 V/div, y-axis: 1 V/div), (b) and (d) phi versus v2 (x-axis: 2 V/div, y-axis: 2 V/div).

3.5 Conclusion

In this chapter, a memristor based modified canonical Chua's circuit is studied. The periodic and hyperchaotic trajectories of the MHC system are shown, and the Lyapunov exponents are calculated to confirm the existence of hyperchaos and chaos. The spectrum of the three maximum Lyapunov exponents indicates that the MHC system exhibits hyperchaos and chaos, and eventually jumps from hyperchaos to period-1 behavior as resistance R increases. The simulation and experimental results both manifest that the proposed circuit can give birth to hyperchaotic or chaotic signals. This circuit is a reliable and palpable resource for generating a wide variety of hyperchaotic signals.

References

1. D.B. Strukov, G.S. Snider, G.R. Stewart and R.S. Williams, "The missing memristor found," *Nature* **453**, 7191, pp. 80–83, (2008).
2. B.C. Bao, Z. Liu and J.P. Xu, "Steady periodic memristor oscillator with transient chaotic behaviours," *Electronics Letters* **46**, pp. 237–238, (2010).
3. D. Batas and H. Fiedler, "A memristor spice implementation and a new approach for magnetic flux controlled memristor modeling," *IEEE Transactions on Nanotechnology* **10**, pp. 250–255, (2011).
4. H. Choi, H. Jung, J. Lee, J. Yoon, J. Park, D.J. Seong, W. Lee, M. Hasan, G.Y. Jung and H.H. Hwang, "An electrically modifiable synapse array of resistive switching memory," *Nanotechnology* **20**, p. 345201, (2009).
5. T. Driscoll, Y.V. Pershin, D.N. Basov and M. Di Ventra, "Chaotic memristor," *Applied Physics A*, DOI 10.1007/s00339-011-6318-z, (2011).
6. Y. Ho, G.M. Huang and P. Li, "Nonvolatile memristor memory: device characteristics and design implications," in *IEEE/ACM 2009 International Conference on Computer-Aided Design (ICCAD)*, pp. 485–490, (2009).

7. Y.V. Pershin and M. Di Ventra, "Experimental demonstration of associative memory with memristive neural networks," (2009) [Online]. Available: http://arXiv.org/abs/arXiv:0905.2935

8. Y.V. Pershin and M. Di Ventra, "Practical approach to programmable analog circuits with memristors," *IEEE Transactions on Circuits and Systems I: Regular Papers* **57**, pp. 1857–1864, (2010).

9. A. Rak and G. Cserey, "Macromodeling of the memristor in SPICE," *IEEE Transactions on Computer-aided Design of Integrated Circuits and Systems* **29**, pp. 632–636, (2010).

10. F.Z. Wang, N. Helian, S. Wu, M.G. Lim, Y. Guo and M.A. Parker, "Delayed switching in memristors and memristive systems," *IEEE Electron Device Letters* **31**, pp. 755–757, (2010).

11. M. Itoh and L.O. Chua, "Memristor oscillators," *International Journal of Bifurcation and Chaos* **18**, pp. 3183–3206, (2008).

12. H.H.C. Iu, D.S. Yu, A.L. Fitch, V. Sreeram and H. Chen, "Controlling chaos in a memristor based circuit using a Twin-T notch filter," *IEEE Transactions on Circuits and Systems I: Regular Papers* **58**, pp. 1337–1344, (2011).

13. B. Muthuswamy and P.P. Kokate, "Memristor-based chaotic circuits," *IETE Technical Review* **26**, pp. 417–429, (2009).

14. B. Muthuswamy, "Implementing memristor based chaotic circuits," *International Journal of Bifurcation and Chaos* **20**, pp. 1335–1350, (2010).

15. B. Muthuswamy and L.O. Chua, "Simplest chaotic circuit," *International Journal of Bifurcation and Chaos* **20**, pp. 1567–1580, (2010).

16. I. Petras, "Fractional-order memristor-based Chua's circuit," *IEEE Transactions on Circuits and Systems II: Express Briefs* **57**, pp. 975–979, (2010).

17. L.O. Chua and G.N. Lin, "Canonical realization of Chua's circuit family," *IEEE Transactions on Circuits and Systems* **37**, pp. 885–902, (1990).

18. S.H. Jo, T. Chang, I. Ebong, B.B. Bhadviya, P. Mazumder and W. Lu, "Nanoscale memristor device as synapse in neuromorphic systems," *NanoLetters* **10**, pp. 1297–1301, (2010).

19. M. Varsace and B. Chandler, "MoNETA: a mind made from memristors," *IEEE Spectrum* **12**, pp. 30–37, (2010).
20. G. Grassi and S. Mascolo, "A system theory approach for designing cryptosystems based on hyperchaos," *IEEE Transactions on Circuits and Systems I: Fundamental Theory and Applications* **46**, pp. 1135–1138, (1999).
21. R. Jia, Q. Huang and J. Peng, "Study of the hyperchaos-based Hash function in e-commerce applications," *International Conference on Intelligent Computation Technology and Automation*, pp. 451–454, (2010).
22. H. Kantz, "A robust method to estimate the maximal Lyapunov exponent of a time series," *Physics Letters A* **185**, pp. 77–87, (1994).
23. A. Wolf, J. Swift, H. Swinney and J. Vastano, "Determining Lyapunov exponents from a time series," *Physica D: Nonlinear Phenomena* **16**, pp. 285–317, (1985).
24. S. Baglio and L. Fortuna, "A singular value decomposition approach to detect chaos in nonlinear circuits and dynamic systems," *IEEE Transactions on Circuits and Systems I: Fundamental Theory and Applications* **41**, pp. 908–912, (1994).
25. L. Dieci, D.R. Robert and S.V. Erik, "On the computation of Lyapunov exponents for continuous dynamical systems," *SIAM Journal on Numerical Analysis* **34**, pp. 402–423, (1997).
26. V. Govorukhin, "Lyapunov exponents for ODEs," http://www.mathsworks.com/matlabcentral/fileexchange/, (2008).

Chapter 4

Realization of an Analog Model of a Memristor Based on a Light Dependent Resistor

In this chapter, an analog model of a memristor using a light dependent resistor (LDR) is presented. This model can be simplified into two parts: a control circuit and a variable resistor. It can be used to easily verify theoretical presumptions about the switching properties of memristors. This LDR based memristor model can also be used in both simulations and experiments for future research into memristor applications. The chapter includes mathematical models, simulations and experimental results.

4.1 Introduction

The memristor from HP labs is not expected to be available in a product until 2013 and there has been no announcement regarding the availability of memristors as two terminal components. Therefore, the design and construction of a functional analog memristor model is essential to further experimental research on memristor applications.

In spite of the rapid progress and achievements that have been made in the application and theoretical simulation models of the memristor, the development of practical and realistic memristor analog simulators is lagging. In 1976, Chua first presented an active memristor simulator circuit to study memristive behavior [1]. In 2010, Valsa *et al.* presented a memristor mimicking circuit [2]. In this circuit, the flux-charge relationship inside the memristor is observable. Although this circuit is excellent for observing memristive behavior, it is not intended to be used for developing memristor applications as the I-V relationship is obtained from observing the currents and the voltages within the control circuit. A plug-and-play kit has also been designed, which mimics the behavior of the HP memristor by producing a pinched loop hysteresis signal [3]. In fact, the HP memristor belongs to the category of charge-controlled memristors according to the theory developed by Chua in 1971 [4].

In this chapter, a memristor analog model using an LDR is described. This model is composed of two parts: a control circuit and a light controlled resistor. The controlling circuit includes three op-amps, eleven resistors and one capacitor. The operational frequency range of this model can be extended to higher frequency ranges by changing the resistor/capacitor relationship in the integrator sub-circuit. The practical upper frequency limit is governed by the physical properties of the LDR. The goal is to separate the circuit that controls the resistance from the variable resistor, so that a signal can both control the resistance and pass through the variable resistor in its original form. This will allow the circuit to be used in practical applications as an actual component.

This chapter is organized as follows: Firstly, a new method for the analog realization of a memristor is proposed and described in detail. Then, the simulation and experimental results are shown to ensure the viability of the model.

4.2 Circuit Objectives

In 2010, Valsa *et al.* presented an analog memristor mimicking circuit [2]. The pinched loop is found by observing the I-V relationship of the input voltage and segmental current inside the circuit. This circuit was designed to effectively observe a memristive response rather than to be used in an application in place of a real memristor.

The control circuit and the voltage-controlled resistor, in this case, a FET sub-circuit, are in series. It was determined the input signal needed to separately control the memristive function and pass in its original form into the memristor mimicking stage. These need to be parallel functions rather than in series. This way the circuit could be used in applications to simulate a functional memristor. These parallel processes can be seen in the block diagram of Fig. 4.2. It also determined the operational frequency range of the circuit that should be as wide as can be reasonably obtained without compromising the behavioral traits.

So, based on the above results and addressing the design goals, a new implementation has been developed and is described in Sec. 4.3.

4.3 Simulation Setup

Here, the goal is to realize a memristor which can be easily used in practical circuits. Based on the flux-control model given in Eqs. (1.6) and (1.7), to find a voltage corresponding to flux (φ), an integrator must be used. Then we will use the voltage corresponding to flux to control the memductance $W(\varphi)$. The integrator stage from the memristor emulating circuit presented in [2] is used as it has been found to perform effectively. Common ways to remotely control a variable resistor include using FET based circuits, circuits based around an operational transconductance amplifier, such as the CA3080, and optocouplers. The LDR/LED optocoupler was determined to be the most suitable method for implementing the design. It is specifically designed for its variable resistance to be controlled remotely, is readily available and maintains simplicity in the design. Within the optocoupler, the LED brightness can be controlled by the input signal and will determine the conductance value of the LDR or photocell. This complies with the goal of separating the control circuit from the varying resistance into parallel functions.

The chosen analog optocoupler is a Silonex NSL-32, consisting of an LED optically coupled to an LDR. The LDR resistance is high when the LED is "off" and low when the LED is "on". The resistance varies according to the brightness of the LED, which in turn is determined by the current flow through the LED. The circuit schematic is shown in Fig. 4.1. From the structure and operating principles, it can be seen that the optocoupler performs as a remotely controlled variable resistor in the circuit. The control circuit is not part of the signal path.

As is shown in Fig. 4.1, the optocoupler has four terminals, two for the LED are connected to the controlling circuit and the other two connections are for the LDR, which is connected into a circuit to emulate a working memristor.

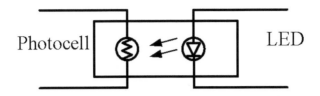

Fig. 4.1. Circuit schematic of Silonex NSL-32 optocoupler.

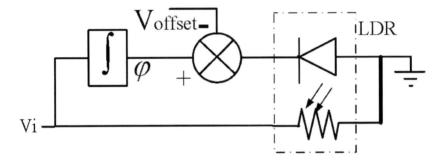

Fig. 4.2. Block diagram of the memristor based on LDR. The upper path is the control circuit. The lower path is the signal path.

Based on the description above, the structure block diagram is shown in Fig. 4.2. V_{offset} is used to offset the signal into a region above the LED turn on voltage and maintain signal processing within the LDR's linear region.

The photocell has a nonlinear characteristic in its response to the intensity of light received from the LED. It is close to linear in the middle region of its operating range and this was determined by placing a 1 k□ current limiting resistor in series with the LED and observing the change in resistance as a voltage was applied to the LED. The circuit presented in this paper also uses a 1 k□ current limiting resistor. The voltage/conductance relationship of the Silonex NSL-32 is shown in Fig. 4.3. By fitting the linear part of the curve, we can determine the working region we want to use. As shown in Fig. 4.3. The region between 4 V and 9 V is nearly linear. Its fitted linear expression is

$$W(v) = 0.3v + 0.11. \tag{4.1}$$

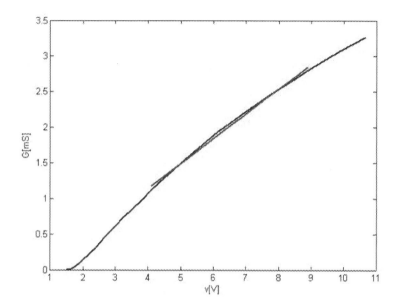

Fig. 4.3. Voltage/conductance characteristics of the Silonex NSL-32 opto-coupler.

In Multisim, a voltage controlled resistor (VCR) model is used to simulate the LDR, based on Eq. (4.1). The complete circuit is given in Fig. 4.4; a description of each stage is outlined below. The first op-amp stage consisting of U1A with R_1 and R_2 forms an inverting buffer to minimize current draw from the input signal. When the input of the memristive system is $v_i(t)$ the output of the op-amp U1A will be $v_2(t)$:

$$v_2(t) = -\frac{R_2}{R_1} v_i(t).$$ (4.2)

R_3, R_4, C_1 and op-amp U1B form an integrator stage, one of the vital elements in the circuit to realize the input voltage as flux (φ), here R_3 and C_1 determine the integration time constant τ. R_4 exists to control DC gain, drift and offset in this stage of the circuit; it has negligible effect during normal operations. The output of the op-amp U1B will be $v_4(t)$:

$$v_4(t) = -\frac{1}{\tau}\int v_2(t)dt = -\frac{1}{R_3 C_1}\int v_2(t)dt = v_\varphi.$$ (4.3)

Realization of a summer with R_5–R_9 and op-amp U1C is shown in front of the LDR, R_6–R_8, VEE and VCC supply V_{offset}, a DC voltage as shown in Fig. 4.2. On the actual experimental circuit R_6 and R_7 are replaced by a multi-turn trimpot. The voltage at point 7 is shown in Eq. (4.4):

$$v_7(t) = -\frac{R_9}{R_5}v_4(t) - \left(\frac{R_9}{R_8}\right) \times V_5$$

$$= -\frac{R_9}{R_5}\left(-\frac{1}{R_3 C_1}\int\left(-\frac{R_2}{R_1}v_i(t)\right)dt\right) - \left(\frac{R_9}{R_5}\right) \times V_5. \qquad (4.4)$$

According to Eq. (4.1), we can determine the memductance of the memristor as below:

$$W(t) = 0.3 \times (0 - v_7(t)) + 0.11. \qquad (4.5)$$

In Fig. 4.4, the diode D_1 and R_{11} protect the LED during set-up and prevent the LED from reverse voltage effects if the R_6/R_7 trimpot is incorrectly adjusted. Once the circuit is set-up to the proper offset these components have no effect. In simulation there is no need to include them.

Using a sinusoidal wave as the input signal we can obtain a value for each point as shown in Eq. (4.6):

$$v_i(t) = V \sin \omega t,$$

$$v_2(t) = -\frac{R_2}{R_1}V \sin \omega t,$$

$$v_4(t) = -\frac{1}{R_3 C_1}\int v_2(t)dt = -\frac{V}{\omega R_3 C_1}\frac{R_2}{R_1}\cos \omega t = v_\varphi,$$

$$v_7(t) = -\frac{R_9}{R_5}v_4(t) - \left(\frac{R_9}{R_8}\right) \times V_5 \qquad (4.6)$$

$$= -\frac{R_9}{R_5}v_\varphi(t) - \left(\frac{R_9}{R_8}\right) \times V_5$$

$$= \frac{R_9 R_2 V}{\omega R_5 R_3 R_1 C_1}\cos \omega t - \frac{R_9}{R_8} \times V_5.$$

From Eq. (4.6), we can see that the voltage v_φ used to control the conductance of LDR in response to the frequency and the amplitude of the input signal, also the value of the memductance can be estimated using Eq. (4.5):

$$W(t) = 0.3 \times \left(0 - \frac{R_9 R_2 V \cos \omega t}{\omega R_5 R_3 R_1 C_1} + \frac{R_9 V_5}{R8} \right) + 0.11 \qquad (4.7)$$

when V is 1 V and ω is chosen to be 2π rad/s, using a LED with the cathode connected to point 7, the voltage is set to approximately +5.5 V, which is close to the center of the LDR's linear region and remains there with an input signal of 1 V amplitude. Then using the parameters shown in Fig. 4.4, we can use Eq. (4.7) to estimate the memductance of the LDR memristor ranges from 1.4522 mS to 2.0678 mS, which means the memristance of the LDR memristor ranges from 483.606 Ω to 688.610 Ω.

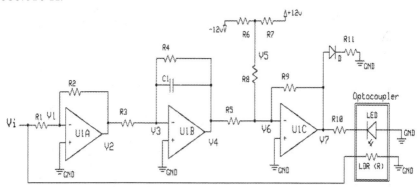

Fig. 4.4. Schematic of the LDR based memristor emulator.

Figures 4.5 and 4.6 show the I-V curves obtained by Multisim simulation with different input frequencies and amplitudes of the input sinusoidal. In Figs. 4.5–4.8 and 4.11–4.13, the y-axis is the current passing through the LDR while the x-axis is the input voltage $v_i(t)$.

It should be noted that the pinched hysteresis loop narrows as higher frequency signals are applied; this is a known behavioral property of memristors [3, 4, 5, 6]. Also, the results with triangle waves and square waves in different frequency are shown in Figs. 4.7 and 4.8.

Table 4.1. Components for the MMC.

Components	Value/Description
R_1, R_2	100 k□
R_6, R_8, R_9	10 k□
R_3	150 k□
R_4	680 k□
R_{10}, R_{11}	1 k□
R_7	4.7k□
R_5	47 k□
C_1	220 nF
C	1 µF
U_1	TL074 quad operational amplifier
Opto-coupler	Silonex NSL-32

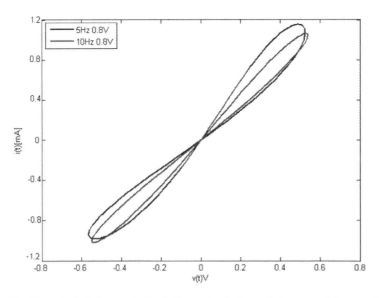

Fig. 4.5. The pinched loop obtained from simulation of the memristor emulator. Characteristics of the LDR memristor for f = 5 Hz, f = 10 Hz for sinusoidal input, amplitude: 0.8 V.

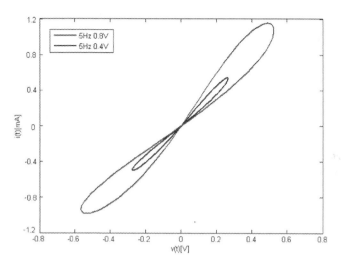

Fig. 4.6. The pinched loop obtained from simulation of the memristor emulator. Characteristics of the LDR memristor for different sinusoidal input amplitudes at f = 5 Hz (0.4 V and 0.8 V).

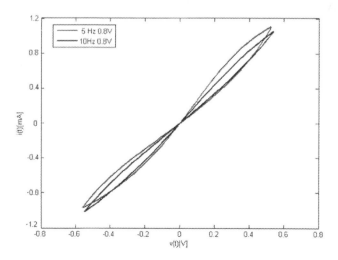

Fig. 4.7. Characteristics of the LDR memristor with triangle wave at input when f = 5 Hz, f = 10 Hz (R3 = 150 k□, C1 = 220 nF) for amplitude 0.8 V.

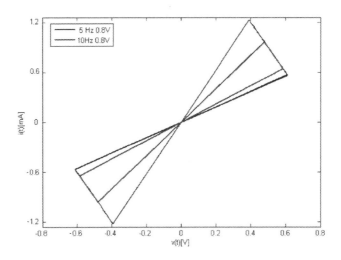

Fig. 4.8. Characteristics of the LDR memristor with square wave at input when f = 5 Hz, f = 10 Hz for amplitude 0.8 V.

4.4 Experimental Setup

Following the successful simulation of the circuit, an LDR based memristor circuit has been constructed. Figure 4.9 shows a Silonex NSL-32 optocoupler. Figure 4.10 shows the constructed circuit of the memristor emulator. The input signal is connected to the input of the control subcircuit and one terminal of the LDR's photocell.

A 500 □ resistor is placed between the other terminal of the photocell and ground. This forms a voltage divider that is used to observe the varying voltage of the signal after it has passed through the memristor emulator. The variable resistance range of the circuit was measured to range between 500 □ and 1 k□ at 1 Hz.

Figure 4.11 shows the experimental results with a sinusoidal input signal of different frequencies and Fig. 4.12 shows the results when the amplitude of the input signal is adjusted. The circuit was found to be stable, easy to set up and consistent in its operation.

Fig. 4.9. The Silonex NSL-32 optocoupler.

Fig. 4.10. Physical implementation of memristor emulator.

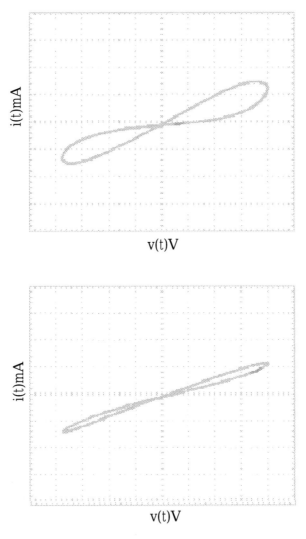

Fig. 4.11. Physical implementation of the memristor emulator: The I-V relationship of LDR memristor at 5 Hz (top) and 10 Hz (bottom) when the sinusoidal input amplitude is 0.8 V. The x-axis scale is 200 mV/division and y-axis scale is 0.4 mA/division.

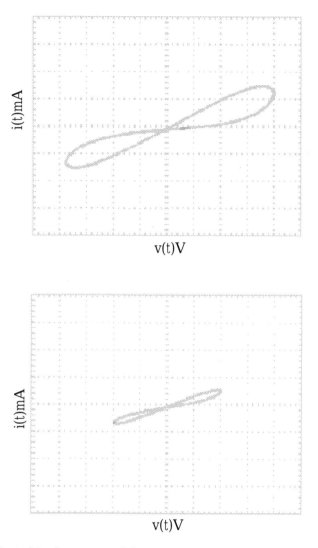

Fig. 4.12. Physical implementation of the memristor emulator: The I-V relationship of LDR memristor for different sinusoidal input amplitudes at f = 5 Hz (0.8 V (top) and 0.4 V (bottom)). The x-axis scale is 200 mV/division and y-axis scale is 0.4 mA/division.

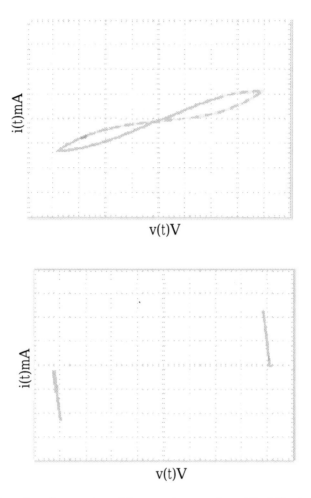

Fig. 4.13. Physical implementation of the memristor emulator: The I-V relationship of LDR memristor with triangle (top) and square input (bottom) signal at 5 Hz when the input amplitude is 0.8 V. The x-axis scale is 200 mV/division and y-axis scale is 0.4 mA/division.

4.5 Conclusion

In this chapter, a memristor simulator using an LDR was presented. After carefully studying the existing theories, the experiment setup was developed. The simulation and experimental results both confirmed that the proposed circuit is capable of simulating a memristor. This memristor simulator is simple to construct and use, it operates effectively in a useful frequency range and, most importantly, is suitable to be used in practical applications and experiments where a memristor is required. The upper frequency range is limited by the characteristics of the LDR in the optocoupler, as the time it takes to vary resistance can be measured in milliseconds.

Optocouplers with faster LDRs are available and these will be examined in the further development of this research. Other types of voltage-controlled and current-controlled resistor circuits will be examined; also methods of implementing a sample and hold based memory function into this circuit are being examined.

References

1. L.O. Chua and S.M. Kang, "Memristive devices and systems," *Proc. IEEE*, vol. 64, pp. 209–223, (1976).
2. J. Valsa, D. Biolek and Z. Biolek, "An analog model of the memristor," *International Journal of Numerical Modeling: Electronic Networks, Devices and Fields*, DOI:10.1002/jnm.786, (2010).
3. A. Sodhi and G. Gandhi, "Circuit mimicking TiO$_2$ memristor: a plug and play kit to understand the fourth passive element," *International Journal of Bifurcation and Chaos*, vol. 20, no. 8, pp. 2537–2545, (2010).
4. L.O. Chua, "Memristor- The missing circuit element," *IEEE Transaction on Circuit Theory*, vol. 18, no. 5, pp. 507–519, (1971).
5. Y.V. Pershin and M. Di Ventra, "Practical approach to programmable analog circuits with memristors," *IEEE Transactions*

on Circuits and Systems-I: Regular Papers, vol. 57, no. 8, pp. 1857–1864, (2010).

6. Z. Kolka, D. Biolek and V. Biolkova, "Hybrid modeling and emulation of mem-systems," *International Journal of Numerical Modeling: Electronic Networks, Devices and Fields*, DOI:10.1002/jnm.825, (2011).

Chapter 5

Design of a Memcapacitor Emulator Based on a Memristor

Building on our previous work to develop an analog model of a memristor, a memcapacitor emulator is proposed in this chapter. This model can be realized by transforming a memristor emulator to a memcapacitor emulator. The characteristics of a memcapacitor are based on the theory proposed by L.O. Chua. The transformation process is described in detail in this chapter which includes simulation and experimental results.

5.1 Introduction

In 2009, two devices that can be considered memristive systems were announced; they have been defined as the memcapacitor and the meminductor [1, 2]. They have similar characteristics to memristors, and their values vary according to the state variables and histories of their systems. Their internal states display obvious memory effects, and the relationships between the two constitutive variables are displayed as hysteretic loops.

Currently there are only a few reports describing analog models of the memcapacitor [3, 4, 5, 6]. In [3], the authors implemented a methodology of SPICE modeling of general memristive, memcapacitative, and meminductive systems on the basis of their mathematical definitions. In [4], the authors proposed a way to realize the mutator of a memcapacitor by utilizing the analog model of memristor. The paper introduced an electronic circuit with memristors that operate as a memcapacitors and meminductors, although the approximate equivalent circuits contained parasitic resistances in series with the simulated elements. In [5, 6], D. Biolek and his colleagues improved on the circuit in [4], and provided a mutator for exact memristor to memcapacitor transformation; in this case the memcapacitor does not include any parasitic elements in its equivalent model.

In Chapter 4, a novel memristor analog model using a light dependent resistor (LDR) was described. This model can be viewed as having two parts; a control circuit which can be seen as a black box and a light controlled resistance. The control circuit consists of three operational amplifier stages and an LED. It has a wide operating frequency, which can be extended into a higher frequency range by changing components in the integrator sub-circuit. Importantly, is suitable to be used in practical applications and experiments where a memristor is required.

In this chapter, by combining the work described in Chapters 4 and 5; a memcapacitor analog model based on the LDR memristor model is presented. This model can be used in both computer simulations and hardware experiments.

This chapter is organized as follows: Firstly, the theory of memcapacitors and parallels to memristive systems is discussed. Secondly, the way to transform a memristor model to a memcapacitor model is described in detail. Finally, the simulation and experimental results are shown to ensure the viability of the model.

5.2 The Memcapacitor and Its Relation to the Memristor

By defining the parameters as follows:

$$\sigma = \int_{t_0}^{t} q(\tau)d\tau, \ \rho = \int_{t_0}^{t} \varphi(\tau)d\tau \tag{5.1}$$

and then combining with (1.1–1.3), we can get the R, L and C elements and their memory versions [7].

Expanding the pinched loop characteristics of memristor's current-voltage relation discussed above, Chua and his co-workers extended the notion of memristive systems to include capacitive and inductive elements [1]. In these two elements, the hysteresis loops are generated from two constitutive variables; charge-voltage and current-flux.

The voltage-controlled memcapacitor is given by (5.2) [1]:

$$q(t) = C_M \left(\int_{t_0}^{t} v_c(\tau) d\tau \right) \cdot v(t) \tag{5.2}$$

and the charge-controlled memcapacitor can be written as:

$$v_c(t) = C_M^{-1} \left(\int_{t_0}^{t} q(\tau) d\tau \right) \cdot q(t). \tag{5.3}$$

Figure 5.2 shows the pinched hysteresis loop of a memcapacitive system. By observing the relationship between the integral of charge σ and the flux φ inside the memcapacitor:

$$C_M = \frac{d\sigma}{d\varphi} \tag{5.4}$$

we can obtain the relationship between the flux φ_1 and the integral of charge σ_1 in the memcapacitor:

$$d\varphi_1 = \frac{1}{C_M} d\sigma_1. \tag{5.5}$$

Subsequently from (1.5), we can obtain the relationship between the flux φ_2 and the charge q_2 in the memristor:

$$d\varphi_2 = R_M dq_2. \tag{5.6}$$

So based on Eqs. (5.5) and (5.6), the relationship of the memristor (MR) and the memcapacitor (MC) is shown in Fig. 5.3. It is clear that the most important condition for transforming the MR model to MC model is the $(\varphi_1, \sigma_1) \Leftrightarrow (\varphi_2, \sigma_2)$ part, which is the core of the MR to MC mutator, by connecting MR to the port 2; we can transform the MR to MC at port 1.

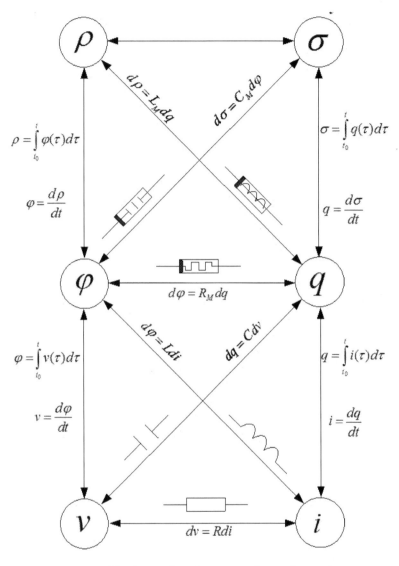

Fig. 5.1. System of fundamental passive memory-less and mem-elements.

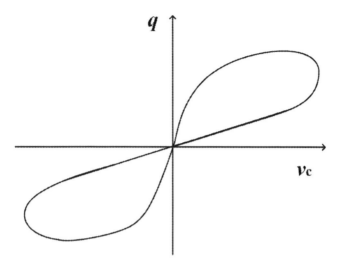

Fig. 5.2. The q-v relationship of a memcapacitor [1].

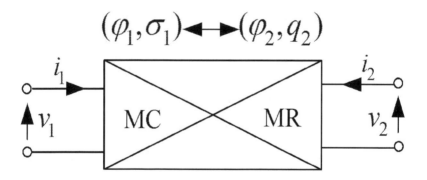

Fig. 5.3. The MR to MC mutator [4].

By observing (5.5), (5.6) and Fig. 5.3, we get an intrinsic relation between memristor and memcapacitor, that is

$$\varphi_1 = k_y v_2$$
$$\sigma_1 = k_x q_2$$

(5.7)

where k_x and k_y are real constants, their values depend on the memcapacitor realization circuit based on the memristor model.

By differentiating (5.7) on both sides (5.8) is obtained:

$$v_1 = k_y v_2$$

$$\frac{1}{k_x}\int_{t_0}^{t} i_1 dt = i_2 \qquad (5.8)$$

where v_1 and i_1, v_2 and i_2 are the voltage and the current in the memristor and memcapacitor circuit respectively.

If we want to realize the memcapacitor mutator we can just utilize the relationship in (5.8) to transform the memristor to memcapacitor.

5.3 Simulation Setup

To realize the relationship between the memristor and memcapacitor, by referencing Chapter 4 and [5] which for the first time proposed the idea of using a memristor to emulate a memcapacitor, the transforming circuit is shown in Fig. 5.4.

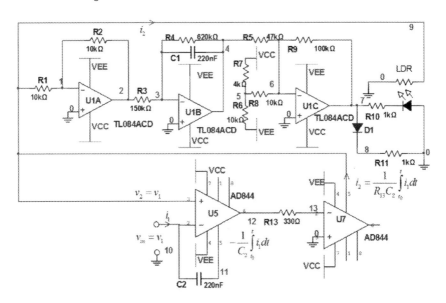

Fig. 5.4. Schematic of the memcapacitor emulator with memristor model using LDR.

In this circuit, the capacitor C2 and U5 are used to integrate the current i_2, the voltage at point 11 is

$$v_{11}(t) = -\frac{1}{C_2} \int_{t_0}^{t} i_1 dt = -\frac{1}{C_2} q(t) \tag{5.9}$$

and the second AD844, U7 and R13 serve to change the voltage to a current, the current flow from pin 5 of U7 is

$$i_2 = \frac{1}{R_{13}C_2} \int_{t_0}^{t} i_1 dt = \frac{1}{k_x} \int_{t_0}^{t} i_1 dt \tag{5.10}$$

where k_x depends on the values of R_{13} and C_2. This is because the current flow into the pin 2 of U7 is

$$i_-(t) = \frac{v_{11}(t) - v_{13}(t)}{R_{13}}$$

$$= \frac{v_{11}(t) - v_-(t)}{R_{13}}$$

$$= \frac{v_{11}(t) - v_+(t)}{R_{13}}$$

$$= \frac{v_{11}(t) - 0}{R_{13}}$$

$$= \frac{1}{R_{13}C_2} \int_{t_0}^{t} i_1 dt \tag{5.11}$$

and the current flow out pin 5 is proportional to the input error current, so

$$i_2 = i_+(t) - i_-(t)$$

$$= 0 - \left(-\frac{1}{R_{13}C_2} \int_{t_0}^{t} i_1 dt \right) \tag{5.12}$$

$$= \frac{1}{R_{13}C_2} \int_{t_0}^{t} i_1 dt .$$

Also in this circuit, we can get the relation of the input voltages between the MR and MC as (5.13):

$$v_1 = v_2 = k_y v_2 \qquad (5.13)$$

where k_y equals to 1. This is realized according to the characteristics of the op-amps.

As we know, the relationship between the current and voltage across a traditional resistor is linear, but the I-V characteristic of the memristor is a pinched loop in response to a sinusoidal input. Similarly, the relationship between the charge and the voltage of the capacitor is linear too, but the curve of q-v of memcapacitor is nonlinear; a hysteresis curve.

The Multisim simulation result of the q-v characteristics of the memcapacitor is shown in Fig. 5.5. The charge is represented by the voltage from point 11, and the voltage is the input voltage at pin 10. Although we want to show the relationship between charge and voltage in the memcapacitor, the voltage at point 11, as shown in (5.9), represents the 'negative image' of the charge. That is why the image is in the second quadrant and the fourth quadrant, instead of in the first quadrant and the third quadrant.

As the memcapacitor emulator is based on the memristor model, if we want the memcapacitor to operate properly, it is essential the memristor emulator works well. As shown in Fig. 5.5, if we want the memcapacitor to work in a desired frequency range, we should adjust the component values in the memristor circuit and then change the component values in the transforming circuit, to make the memcapacitor work in the same range. This is why we give the values of four parameters (R3, C1, R13 and C2) for each figure.

Also, from studying the relationship between the memristor and the memcapacitor, we can get the memcapacitance from the transforming circuit as below:

$$C_M = \frac{d\sigma_1}{d\varphi_1} = \frac{k_x q_2}{k_y \varphi_2} = \frac{k_x}{k_y R_M} = \frac{R_{13} C_2}{R_M} R_{13} C_2 W_M . \qquad (5.14)$$

When the input voltage is $v_i(t)$, the memductance $W_M(v_\varphi)$ is calculated by:

$$W_M(v_\varphi) = 0.3 \times \left(0 - \left(\frac{R_9}{R_5} v_\varphi(t) - \left(\frac{R_9}{R_8} \right) \times V_5 \right) \right) + 0.11 \qquad (5.15)$$

and

$$v_\varphi(t) = v_4(t) = -\frac{1}{R_3 C_1} \int \left(-\frac{R_2}{R_1} v_i(t) \right) dt . \qquad (5.16)$$

So we obtain (5.17):

$$W_M(t) = 0.3 \times \left(0 - \left(-\frac{R_9}{R_5} \left(-\frac{1}{R_3 C_1} \int \left(-\frac{R_2}{R_1} v_i(t) \right) dt \right) \right. \right.$$
$$\left. \left. - \left(\frac{R_9}{R_8} \right) \times V_5 \right) \right) + 0.11 . \qquad (5.17)$$

And according to (5.14) we can evaluate the value of the memcapacitor as:

$$C_M(t) = R_{13} C_2 \left(0.3 \times R_9 \left(\frac{1}{R_5 R_3 C_1} \int \left(-\frac{R_2}{R_1} v_i(t) \right) dt \right) \right.$$
$$\left. + \left(\frac{1}{R_8} \right) \times V_5 \right) + 0.11 . \qquad (5.18)$$

In the special case, when $v_i(t) = V \sin \omega t$, where V is 1 and ω is chosen to 2π rad/s, the voltage V_5 is chosen to equal +5.5 V and using the parameters in the Fig. 5.4, we can estimate the memristance range of the circuit is from 544 Ω to 831 Ω. So the range of the memcapacitor will be 87.30 nF to 133.37 nF.

Figures 5.5 and 5.6 show the q-v curves obtained by Multisim simulation with different input frequencies and amplitudes of the input sinusoidal, but it should be noted that q(t) is $-1/C_2$ times itself. The

Development of Memristor Based Circuits

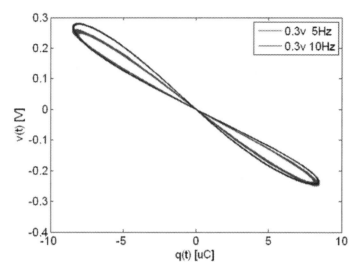

Fig. 5.5. The pinched loop obtained from simulation of the memcapacitor emulator: Characteristics of the memcapacitor for f = 5 Hz (R3 = 100 kΩ, C1 = 200 nF, R13 = 330 Ω, C2 = 220 nF), f = 10 Hz (R3 = 50 kΩ, C1 = 50 nF, R13 = 330 Ω, C2 = 220 nF) when sinusoidal input amplitude is 0.3 V.

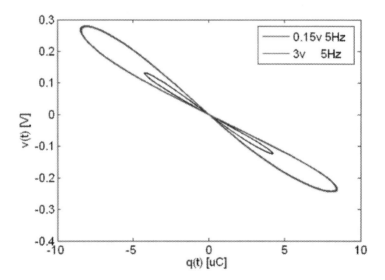

Fig. 5.6. The pinched loop obtained from simulation of the memcapacitor emulator: Characteristics of the memcapacitor for different sinusoidal input amplitudes at f = 5 Hz (0.15 V and 0.3 V) with R3 = 150 kΩ, C1 = 220 nF, R13 = 330 Ω, C2 = 220 nF.

pinched hysteresis loop narrows as higher frequency signals are applied; this is a known behavioral property of memristors. From these figures, using the analog model described in this paper, we can observe the ideal pinched loop from 0.1 Hz to 280 Hz by changing the values of R3, C1, R13 and C2. As described, this is necessary to obtain authentic memristive behavior over a wide frequency range to enable implementation in different functional circuits

5.4　Experimental Setup

Following the successful simulation of the transforming circuit, the actual circuit has been constructed and tested, based on the memristor emulator using LDR in Chapter 4.

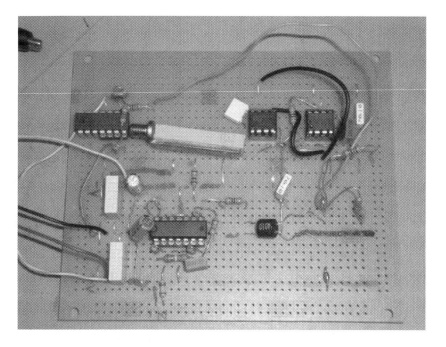

Fig. 5.7. Physical implementation of memcapacitor emulator based on the analog model of memristor using LDR.

In Chapter 4, the variable resistance range of the memristor circuit, which is approximately 500 Ω to 1 kΩ at 1 Hz when the voltage across the 1 kΩ resistor in series with the LDR from +4 V to +7 V. Based on (5.17), we can estimate the range of the value of the memcapacitor, its value is 72.6 nF to 145.2 nF when working at 1 Hz.

The experimental results with sinusoidal input signals are shown in Figs. 5.8 to 5.11.

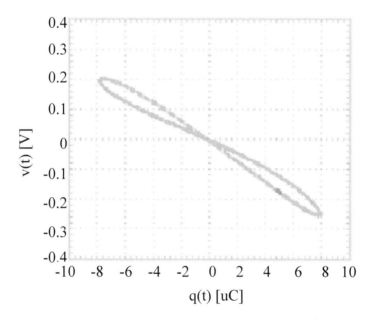

Fig. 5.8. Physical implementation of the memcapacitor emulator based on the LDR memristor. The (-q)-v relationship of memcapacitor at 5 Hz when the sinusoidal input amplitude is 0.4 V. The x-axis and y-axis scales are 100 mV/division and 2 V/division. The x-axis is the charge at the point 11 in the mimicking circuit and the y-axis is the voltage drop on the memcapacitor.

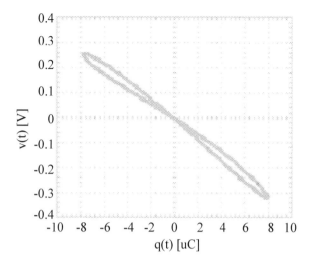

Fig. 5.9. Physical implementation of the memcapacitor emulator based on the LDR memristor. The (-q)-v relationship of memcapacitor at 10 Hz when the sinusoidal input amplitude is 0.4 V. The *x*-axis and *y*-axis scales are 100 mV/division and 2 V/division. The *x*-axis is the charge at the point 11 in the mimicking circuit and the *y*-axis is the voltage drop on the memcapacitor.

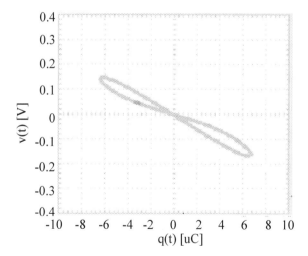

Fig. 5.10. Physical implementation of the memcapacitor emulator based on the LDR memristor. The (-q)-v relationship of the memcapacitor when the sinusoidal input amplitude is 0.4 V at *f* = 5 Hz, the *x*-axis and *y*-axis scales are 100 mV/division and 2 V/division. The *x*-axis is the charge at the point 11 in the mimicking circuit and the *y*-axis is the voltage drop on the memcapacitor.

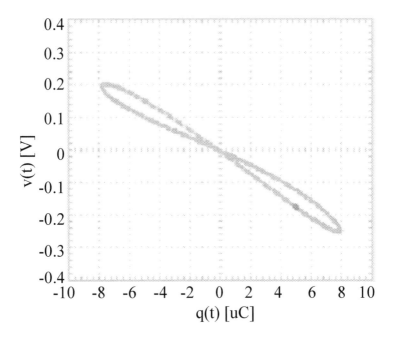

Fig. 5.11. Physical implementation of the memcapacitor emulator based on the LDR memristor. The (-q)-v relationship of the memcapacitor when the sinusoidal input amplitude is 0.8 V at 5 Hz. The x-axis and y-axis scales are 200 mV/division, the *x*-axis is the charge at the point 11 in the mimicking circuit and the *y*-axis is the voltage drop on the memcapacitor.

5.5 Conclusion

In this chapter, a memcapacitor simulator based on an LDR memristor is presented. After carefully studying the existing theories, the experiment setup is presented. The simulation and experimental results both manifest that the proposed circuit is capable of simulating the signal response characteristics of a memcapacitor. As previously noted the circuit is unable to implement the ability to memorize as an actual memcapacitor can be expected to do. Further research is addressing this issue.

References

1. M. Di Ventra, Y.V. Pershin and L.O. Chua, "Circuit elements with memory:memristors, memcapacitors and meminductors," *Proceedings of the IEEE*, vol. 97, no. 10, pp.1717–1724, (2009).
2. M. Di Ventra, Y.V. Pershin and L.O. Chua, "Putting memory into circuit elements:memristors, memcapacitors and meminductors," *Proceedings of the IEEE*, vol. 97, no. 8, pp.1371–1372, (2009).
3. D. Biolek, Z. Biolek and V. Biolkova. "Spice modeling of memcapacitor," *Electronics Letters*, vol. 46, no, 7, pp. 520–522, (2010).
4. Y.V. Pershin and M. Di Ventra, "Memristive circuits simulate memcapacitors and meminductors," *Electronics Letters*, vol. 46, no. 7, pp. 517–518, (2010).
5. D. Biolek and V. Biolkova, "Mutator for transforming memristor into memcapacitor," *Electronics Letters*, vol. 46, no. 21, pp. 1428–1429, (2010).
6. D. Biolek, V. Biolkova and Z. Kolka, "Mutators simulating memcapacitors and meminductors," in *Proceedings of 2010 Asia Pacific Conference on Circuits and Systems (APPCAS 2010)*, pp. 800–803, (2010).
7. D. Biolek, Z. Biolek and V. Biolkova, "SPICE modeling of memristive, memcapacitive and meminductive systems," in *2009 European Conf. on Circuit Theory and Design*, pp. 249–252, (2009).

Chapter 6

Practical Realization of an Analog Model of a Memcapacitor

Although not yet realized as a physical component, the memcapacitor has the potential to be another important memory circuit element due to its energy storage capability. A practical circuit that operates approximately as a memcapacitor is constructed and its operating characteristics are examined by experiment. The circuit is intended to be used as a substitute component in experimental designs where memcapacitors are required.

6.1 Introduction

In 2009, Di Ventra *et al.* proposed the logical development of memristive systems should include capacitive and inductive devices, named memcapacitors and meminductors [1]. As per memristors, a hysteretic loop is observed in the relationship of their characteristic variables, being charge/voltage for memcapacitors and flux/current for meminductors. This proposal motivated a new direction of research into memcapacitative devices [2–4]. Further work describes possible methods to realise a solid state memcapacitive system by modifying existing capacitor designs [5–7]. The impact of memcapacitors as storage elements in all forms of computing; analog, digital and neuromorphic is expected to be huge.

Currently, memcapacitors can only be studied by simulation as a practical physical component has yet to be developed. A similar situation exists in regards to memristors. They are not yet available and emulating circuits must be used. A number of memristor mimicking circuits

(MMC) have been proposed [8–10]. One of the major considerations is whether the emulator serves as a means to simply observe the definitive hysteretic loop as a response to a varying input signal or the emulator can be used in a practical circuit as a functional memristor. The light dependent resistor (LDR) based design proposed in [9] is capable of being used as a circuit component which makes it an attractive choice to use as the basis for a memcapacitor emulator.

In this chapter, we show how to extend the LDR memristor to a memcapacitor emulator by using easily obtainable and inexpensive components. The memcapacitor emulator can be obtained by connecting the proposed MMC [11] to a capacitor multiplier [6].

This chapter is organized as follows: Firstly a description of the proposed circuit is given in sections, and then the complete circuit is synthesized. Secondly, the experiment setup is explained and results are given. Thirdly, the results are evaluated and the viability of the design is discussed.

6.2 Circuit Description

The circuit can be discussed by first looking at the two sub-circuits; the MMC and the capacitor multiplier, then considering the complete circuit.

6.2.1 *Memristor mimicking circuit*

The memristor mimicking circuit (MMC) is comprised of 3 operational amplifier stages and an analog opto-coupler, which contains an LDR and an LED. The input signal takes two paths through the MMC. One path is directly into the LDR, the other path is into the control circuit that determines the resistance value of the LDR. The first operational amplifier stage is a difference inverting buffer that can be modified to adjust the gain of the input signal if necessary. The integrated value of the signal is obtained from the second stage. This is a voltage corresponding to flux (φ) which will be used to remotely control the resistance value of the LDR. This varying, controlled resistance is the emulation of the memristive function. The third operational amplifier stage serves to introduce an offset voltage and drives the LED that controls the LDR by its brightness. The offset voltage ensures the LED is always on.

The block diagram of this circuit is shown in Fig. 6.1, this version of the circuit as shown and described assumes the output terminal is connected to ground. The full MMC with grounded output terminal is given in Fig. 6.2.

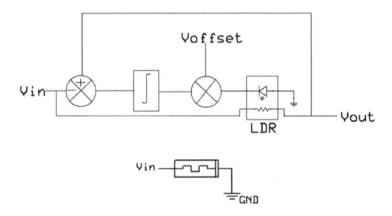

Fig. 6.1. Block diagram of the memristor mimicking circuit from [9] and its equivalent circuit (bottom).

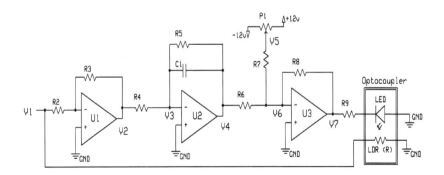

Fig. 6.2. Memristor emulating circuit [9]. See Table 6.1 for component values.

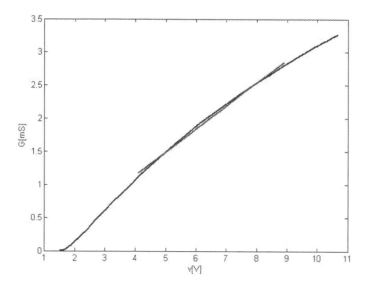

Fig 6.3. Voltage/conductance characteristics of the Silonex NSL-32 opto-coupler [9].

To determine how the resistance of the LDR varies in response to a sinusoidal input, we can consider the equations given in [9] then modify them to suit the current circuit and use the result to determine the varying capacitance in response to the input signal. The following equations describe the MMC where the output is grounded.

The first step is to determine the approximate linear operating region of the LDR as a relationship of the voltage across the LED, through a 1 kΩ resistor (R9), and the conductance of the LDR. This data is used to set the offset voltage of the MMC to the center of the linear region.

Figure 6.3 shows the approximate linear region is between 4 V and 9 V, giving a fitted linear expression of

$$W(v) = 0.3v + 0.11 \qquad (6.1)$$

where W is conductance (mS) and v is voltage (V). Equation (6.1) only describes the characteristics of the LDR as a stand-alone component.

The MMC is examined and defined using Eqs. (6.1)–(6.7). It is important to understand the characteristics of the MMC sub-circuit before examining the complete memcapacitor emulator. For input $v_1(t)$

$$v_2(t) = -\frac{R_3}{R_2}v_1(t).$$ (6.2)

The integrator stage at U_2 realizes $v_4(t)$ as flux (φ), R_4 and C_1 establish the integration time constant τ.

$$v_4(t) = -\frac{1}{\tau}\int v_2(t)\,dt = -\frac{1}{R_4 C_1}\int v_2(t)\,dt = v_\varphi.$$ (6.3)

The next stage, U_3, sums v_φ and the offset voltage v_5.

$$v_7(t) = -\frac{R_8}{R_6}v_4(t) - \frac{R_8}{R_7}v_5.$$ (6.4)

We can see that $v_7(t)$ can be obtained by substituting (6.2) into (6.3) and then (6.3) into (6.4). We can then insert (6.4) into (6.1) to obtain an equation describing the memductance $W(t)$ of the MMC.

$$W(t) = 0.3(0 - v_7(t)) + 0.11.$$ (6.5)

Using a sinusoidal signal at the input gives:

$$v_1(t) = V\sin(\omega t).$$ (6.6)

Allowing $v_7(t)$ to be determined as:

$$v_7(t) = \frac{R_8 R_3 V}{R_2 R_4 R_6 C_1 \omega}\cos(\omega t) - \frac{R_8}{R_7}v_5.$$ (6.7)

For V of 1 V, ω = 2π rad/s, offset voltage V_5 = 5.5 V and the component values given for the schematic in Table 6.1, (6.5) and (6.7) can be used to determine that the memductance of the MMC varies between 1.45 mS and 2.07 mS. This gives a memristance range of 483.1 Ω to 689.7 Ω.

This range can be adjusted by a number of methods. The gain of the first operational amplifier stage at U_1 can be altered to give a corresponding change in the value of $v_7(t)$. This can also be adjusted at the summing stage U_3 by changing the values of components R_8 or R_6.

6.2.2 *Capacitor multiplier*

The capacitor multiplier is used to reduce ripple voltage in amplifiers where circuit board space and component costs are major considerations.

There are two drawbacks of this circuit. One terminal of the capacitor must be connected to ground and the resistor R_1 remains in series with Cm. It is preferable to select R_1 to be as small as practical to reduce the parasitic resistance.

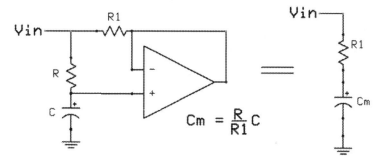

Fig. 6.4. Capacitor multiplier and its equivalent circuit.

6.2.3 *Memcapacitor emulator*

In order to create a memcapacitor emulator, the resistor R in Fig. 6.4 is replaced with the LDR seen in the block diagram of Fig. 6.1. The resistive value of the LDR corresponds to the value of the flux obtained from integrating the varying voltage of the input signal. So when the LDR is used as resistor R in the capacitor multiplier circuit, it can be seen that the value of C_M will also be varied according to the relationship [6]:

$$C_M = \frac{R_M}{R_1} C .$$
(6.8)

The equivalent circuit of the MMC and the capacitor multiplier circuit are combined in Fig. 6.5. Components C and R_1 are from the circuit in Fig. 6.4. C_M is the equivalent capacitance value of the complete circuit and R_M is the memristance value obtained from the MMC.

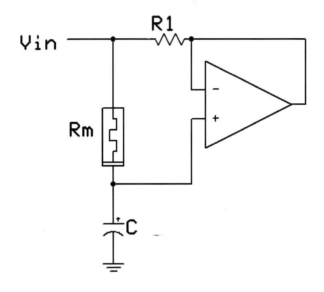

Fig. 6.5. Memristor based capacitance multiplier.

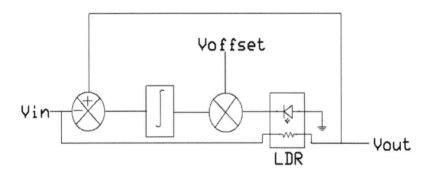

Fig. 6.6. Modified memristor mimicking circuit block diagram.

The MMC from Fitch *et al.* must be modified to be used in the memcapacitor emulator [9]. In the original circuit, as one terminal of the MMC is tied to ground, the signal into the integrator could be taken as

$$V_{in} - 0 = V_{in} . \tag{6.9}$$

Figure 6.6 shows how the circuit is altered to allow for $V_{out} \neq 0$. In the memcapacitor emulator, the signal passed into the integrator is the voltage drop across the MMC, in terms of the circuit itself this is the voltage drop across the LDR. To achieve this, the first operational amplifier stage, U_1, is converted to a differential amplifier.

The complete circuit is given in Fig. 6.7. Component values are given in Table 6.1.

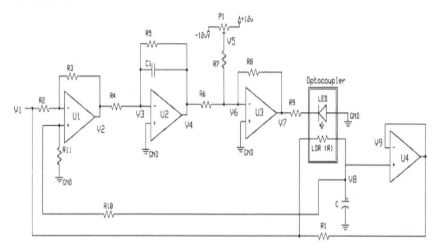

Fig. 6.7. Memcapacitor emulator circuit.

Table 6.1. Components for memcapacitor circuit.

Components	Value/Description
R_2, R_3, R_{10}, R_{11}	100 kΩ
R_6, R_7, R_8	10 kΩ
R_4	150 kΩ
R_5	680 kΩ
R_9	1 kΩ
R_1	22 Ω
P_1	100 kΩ potentiometer
C_1	220 nF
C	1 µF
U_{1-4}	TL074 quad operational amplifier
Opto-coupler	Silonex NSL-32

The values of resistor R_1 and capacitor C can be chosen to suit the desired value and varying range of the simulated memcapacitance. The circuit used to obtain the experimental results given in this paper contained the component values of R_1 = 22 □ and C = 1 µF electrolytic. The opto-coupler is a Silonex NSL-32.

Applying the above values for C and R_1 with the calculated maximum and minimum values for R_M from (6.8), we are able to obtain an expected capacitance varying from 22.0 µF to 31.3 µF.

The capacitance multiplier coupled with the MMC gives a time dependent capacitance $C_M(t)$ that responds to the value of $R_M(t)$. The equations describing the relationship are based on those given by Pershin and Di Ventra [8]. So from (6.8), we obtain:

$$C_M(t) = \frac{R_M(t)C}{R_1}.$$ (6.10)

6.3 Experimental Setup

In order to observe the effectiveness of the circuit as a memcapacitor emulator, the relationship between voltage and charge must be considered. The voltage to be observed is the voltage drop across the LDR; this can be observed at the output of the first operational amplifier differential stage.

The parasitic resistor R_1 can be used to obtain a value representing charge. The voltage across R_1 represents the current passing through the memcapacitor when R_1 << R_M. To observe a value representing charge (*q*) it is necessary to integrate this voltage. This representative value is obtained with a voltage difference stage.

The sub-circuit used to obtain a voltage representing charge is shown in Fig. 6.8. Component values are given in Table 6.2.

The circuit is very similar to the first two stages of the MMC except the differential stage has 10X gain. The resistance value of R_1 is quite small so the voltage drop across it is correspondingly small; gain is introduced to obtain reasonably clean signals for the oscilloscope. This sub-circuit is only used to enable observation of the *q-v* relationship. The circuit was supplied with 1 V p-p and 2.5 V p-p sinusoidal signals.

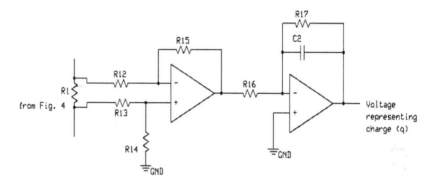

Fig. 6.8. Sub-circuit to obtain a voltage representing charge (q).

Table 6.2. Components for sub-circuit in Fig. 6.8.

Components	Value/Description
R12, R13	100 kΩ
R14, R15	1 MΩ
R16	150 kΩ
R17	680 kΩ
C2	220 nF
U5 - 6	TL072 operational amplifier

The experimental results shown in Figs. 6.9–6.11 display the hysteretic loops that confirm the circuit behaves as a memcapacitive system. Figure 6.9 shows when the frequency is 2 Hz. The plots in Fig. 6.10 show the loop has narrowed at the higher frequency to give an approximate linear relationship between charge and the voltage across the capacitor. This is a normal characteristic of memristive systems. Figure 6.11 shows the hysteretic loop still narrows and starts to curve and become offset at 10 Hz. This distortion is due to the physical characteristics of the LDR. The effect can be eliminated to some extent by adjusting the offset voltage for the LED.

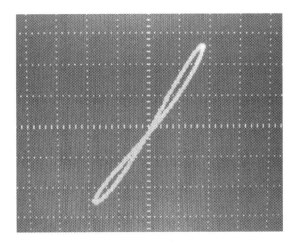

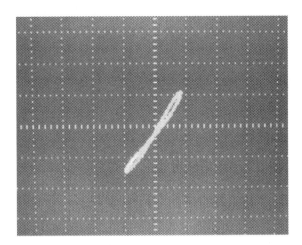

Fig. 6.9. q-v relationship of the memcapacitor emulator circuit at 2 Hz. Input signal: sinusoidal 2.5 V p-p (top) 1.0 V p-p (bottom). y-axis: charge (q) 200 mV/div. x-axis: voltage across capacitor C 200 mV/div.

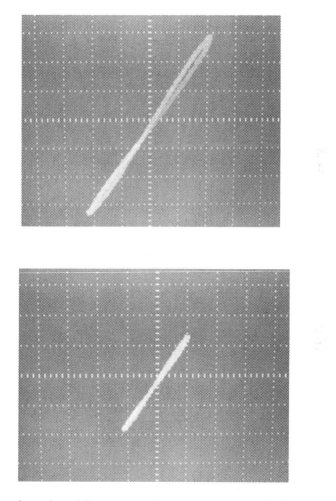

Fig. 6.10. q-v relationship of the memcapacitor emulator circuit at 5 Hz. Input signal: sinusoidal 2.5 V p-p (top) 1.0 V p-p (bottom). y-axis: charge (q) 200 mV/div. x-axis: voltage across capacitor C 200 mV/div.

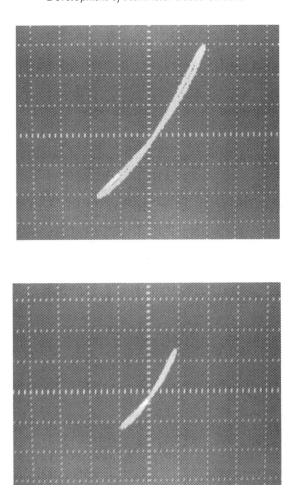

Fig. 6.11. q-v relationship of the memcapacitor emulator circuit at 10 Hz. Input signal: sinusoidal 2.5 V p-p (top) 1.0 V p-p (bottom). y-axis: charge (q) 200 mV/div. x-axis: voltage across capacitor C 200 mV/div.

6.4 Conclusion

The memcapacitor emulator circuit presented has been shown to work well and can be used as an inexpensive and simple-to-build means of studying memcapacitive behavior in experimental circuits. Further research will attempt to improve performance at higher frequencies by substituting faster opto-couplers and implementation in practical circuits to gauge its suitability as a research tool.

References

1. M. Di Ventra, Y.V. Pershin and L.O. Chua, "Circuit elements with memory: memristors, memcapacitors and meminductors," *Proceedings of the IEEE*, vol. 97, no. 10, pp. 1717–1724, (2009).
2. L.O. Chua and S.M. Kang, "Memristive devices and systems," *Proceedings of the IEEE*, vol. 64, no. 2, pp. 209–223, (1976).
3. D. Biolek, Z. Biolek and V. Biolkova, "SPICE modeling of memristive, memcapacitative and meminductive systems," *European Conference on Circuit Theory and Design*, ECCTD 2009, pp. 249–252, (2009).
4. J. Valsa, D. Biolek and Z. Biolek, "An analog model of the memristor," *International Journal of Numerical Modeling: Electronic Networks, Devices and Fields*, vol. 24, pp. 400–408, (2010).
5. D. Biolek and V. Biolkova, "Mutator for transforming memristor into memcapacitor," *Electronics Letters*, vol. 46, no. 21, pp. 1428–1429, (2010).
6. D. Biolek, V. Biolkova and Z. Kolka, "Mutators simulating memcapacitors and meminductors," *2010 IEEE Asia Pacific Conference on Circuits and Systems* (APCCAS), pp. 800–803, (2010).
7. J. Martinez-Rincon, M. Di Ventra, and Y.V. Pershin, "Solid-state memcapacitive system with negative and diverging capacitance," *Phys. Rev. B*, vol. 81, no. 19, 195430, (2010).

8. Y.V. Pershin and M. Di Ventra, "Memristive circuits simulate memcapacitors and meminductors," *Electronics Letters*, vol. 46, no. 7, pp. 517–518, (2010)

9. A.L. Fitch, H.H.C. Iu, X.Y. Wang, V. Sreeram and W.G. Qi, "Realization of an analog model of memristor based on light dependent resistor," *2012 IEEE International Symposium on Circuits and Systems*, accepted.

10. Y.V. Pershin and M. Di Ventra, "Practical approach to programmable analog circuits with memristors," *IEEE Transactions on Circuits and Systems-I: Regular Papers*, vol. 57, no. 8, pp. 1857–1864, (2010).

Chapter 7

Chaos in Memristively Coupled Harmonic Oscillators

This chapter examines a circuit consisting of two linear harmonic oscillators coupled by a memristor. The circuit is chaotic at many settings and this complexity helps to provide a better understanding of the characteristics and properties of memristors.

7.1 Introduction

Research into coupled oscillators is a very old but constantly active field that continues to produce a rich and novel supply of data. This is in part due to the analogies of coupled oscillators across numerous physical systems such as biological processes, electronic circuits, springs and pendulums. This chapter describes a novel circuit that uses a memristor as the resistive coupling element in a pair of coupled harmonic oscillators. The observation of this system can be expected to provide further understanding of the properties of memristors and of coupled oscillators. The memristor mimicking circuit is based on that given by Muthuswamy and Chua in [1] and two sets of capacitors and inductors are added to form the pair of harmonic oscillators.

This chapter will briefly consider couple oscillators, then the memristor mimicking circuit (MMC) used in this system. Section 7.3 will examine the actual memristively coupled harmonic oscillator circuit being discussed and then Sec. 7.4 describes the experimental setup and results obtained.

7.2 Coupled Oscillator Circuits

There have been many studies on coupled oscillators, generally focusing on the synchronous, asynchronous or chaotic phenomena observed in these circuits [2–8]. These types of circuits find applications in, amongst other fields, neural networks, wireless systems, secure communications and image processing [4–7]. Much work in this field focuses on synchronization phenomena and uses in cellular neural networks [9–13].

Usually oscillators are coupled by a single resistor to ground, although there are many alternative ways for coupling to be arranged. The synchronization phenomena, or synchrony, observed is known to be dependent on the types of oscillators, Wien or van der Pol for example, and generally manifests as in-phase, anti-phase or N-phase (where N is the number of oscillators in the circuit) [9].

7.3 Memristor Mimicking Circuit

The MMC used for the memristively coupled harmonic oscillators, described in this chapter, is from the design presented by Muthuswamy and Chua in [1] with changes to the resistors determining the values of α and β. In [1], this memristor mimicking circuit serves to emulate a memristor in a chaotic circuit consisting of just three components, two being linear and passive; a capacitor and an inductor, the third being nonlinear and active; a memristor.

This MMC is actually a memristive system based on the description given by Chua and Kang in [14] rather than the ideal version first published by Chua [1, 15] in 1971. The equations describing this system are given in [1] as

$$V_M = \beta(x^2 - 1)i_M, \tag{7.1}$$

$$\dot{x} = i_M - \alpha x - i_M x. \tag{7.2}$$

For the circuit:

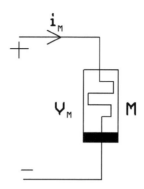

Fig. 7.1. Circuit to show memristor function [1].

Figure 7.2 is a plot of the memristive function, shown in Eq. (7.3), used in [1] being the definition of memristance for this system

$$R(x) \triangleq \beta(x^2 - 1).$$ (7.3)

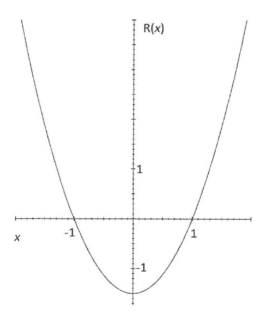

Fig. 7.2. Memristance as a function of internal state [1].

The circuit of the MMC is shown in Fig. 7.3; this is equivalent to the circuit of Fig. 7.4.

In [1], the value for R17 (β) is given as a 5 k☐ potentiometer and for R16 (α) is given as a 10 k☐ potentiometer.

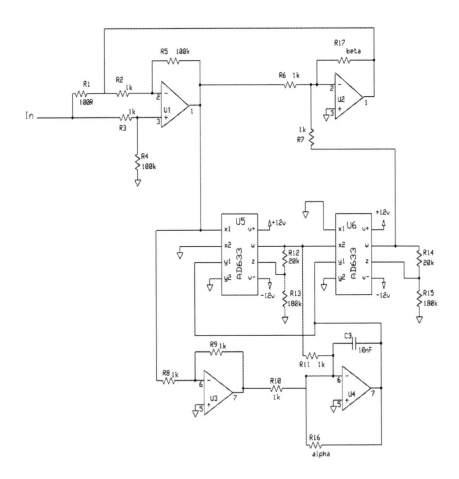

Fig. 7.3. Memristor mimicking circuit [1].

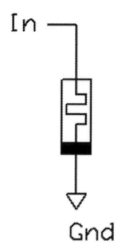

Fig. 7.4. The MMC is equivalent to a memristor with one terminal grounded.

7.4 Memristively Coupled Harmonic Oscillator Circuit

It can be seen that the circuit is an extension of the simplest chaotic circuit presented in [1], where L_2 and C_2 are added to form 2/3 the second harmonic oscillator. The MMC requires changes to the values of α and β, but is otherwise the same. The circuit equations are

$$\dot{v}_1 = \frac{i_1}{C_1}, \tag{7.4}$$

$$\dot{i}_1 = \frac{-1}{L_2}\left(v_1 + R(x_m)(i_1 - i_2)\right), \tag{7.5}$$

$$\dot{v}_2 = \frac{i_2}{C_2}, \tag{7.6}$$

$$\dot{i}_2 = \frac{1}{L_2}\left(-v_2 + R(x_m)(i_1 - i_2)\right), \tag{7.7}$$

$$\dot{x}_m \triangleq -(i_1 - i_2) - \alpha x_m + (i_1 - i_2)x_m, \tag{7.8}$$

$$R(x_m) = \beta(x_m{}^2 - 1). \tag{7.9}$$

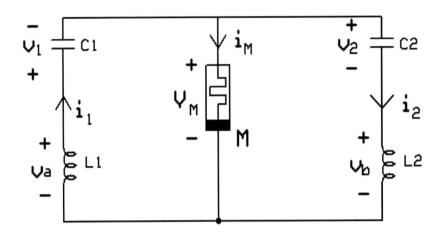

Fig. 7.5. Memristively coupled harmonic oscillators.

7.5 Experimental Setup

The circuit was constructed with the MMC as per Fig. 7.3 and the harmonic oscillators as per Fig. 7.5. The values for the components of the harmonic oscillator are given in tables for each set of images. Potentiometers were used for R1, β and α. This was done to allow the circuit behavior to be observed for a wide range of values.

Table 7.1. Component values used to obtain signals as shown in Figs. 7.6–7.10.

Component	Value
L1	10 mH
L2	40 mH
C1	290 pF
C2	220 pF
R_1	15 Ω
R_α	1.7 MΩ

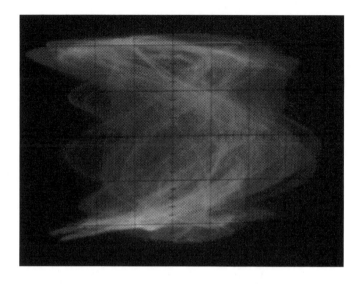

Fig. 7.6. Observations of the voltages at the L1/C1 and L2/C2 nodes for $R\beta$ = 1.7 kΩ. The oscilloscope is set in X-Y mode at 5 V/div.

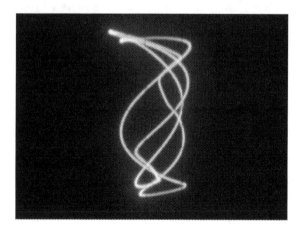

Fig. 7.7. Observations of the voltages at the L1/C1 and L2/C2 nodes for $R\beta$ = 3.0 kΩ. The oscilloscope is set in X-Y mode at 5 V/div.

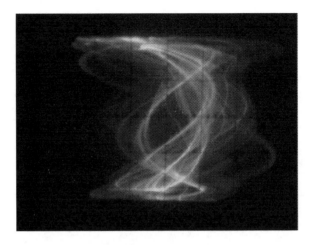

Fig. 7.8. Observations of the voltages at the L1/C1 and L2/C2 nodes for Rβ = 3.7 kΩ. The oscilloscope is set in X-Y mode at 5 V/div.

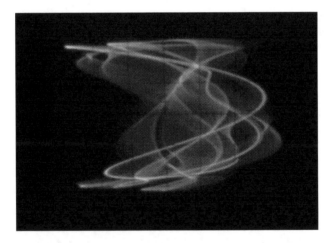

Fig. 7.9. Observations of the voltages at the L1/C1 and L2/C2 nodes for Rβ = 10.2 kΩ. The oscilloscope is set in X-Y mode at 5 V/div.

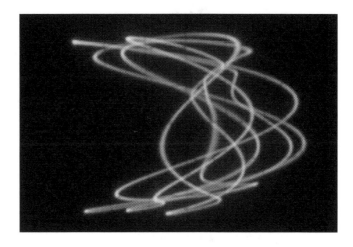

Fig. 7.10. Observations of the voltages at the L1/C1 and L2/C2 nodes for Rβ = 10.5 kΩ. The oscilloscope is set in X-Y mode at 5 V/div.

Fig. 7.11. Observations of the voltages at the L1/C1 and L2/C2 nodes for Rβ = 3.3 kΩ. The oscilloscope is set in X-Y mode at 5 V/div.

Fig. 7.12. Observations of the voltages at the L1/C1 and L2/C2 nodes for $R\beta$ = 5.5 kΩ. The oscilloscope is set in X-Y mode at 5 V/div.

Table 7.2. Component values for Figs. 7.11 & 7.12.

Component	Value
L1	10 mH
L2	40 mH
C1	470 pF
C2	220 pF
R_1	100 Ω
R_α	1.7M Ω

Table 7.3. Component values for Figs. 7.13 & 7.14.

Component	Value
L1	10 mH
L2	40 mH
C1	470 pF
C2	33 pF
R_1	200 Ω
R_α	1.7 MΩ

The circuit was found to generate chaotic and complex signals for many values of α, β, C_1 and C_2 with periodic episodes or no oscillation at other settings.

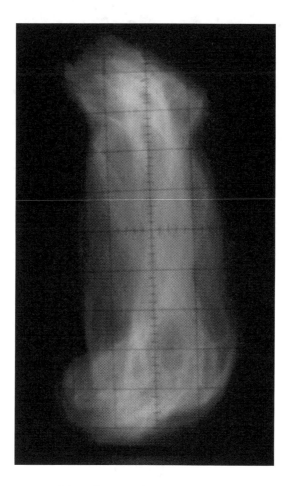

Fig. 7.13. Observations of the voltages at the L1/C1 and L2/C2 nodes for $R\beta$ = 1.5 kΩ. The oscilloscope is set in X-Y mode at 5 V/div.

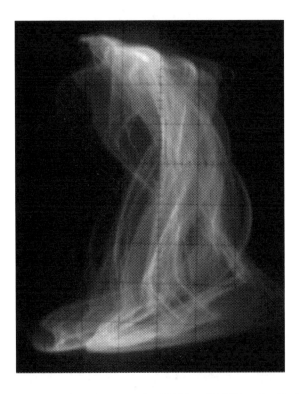

Fig. 7.14. Observations of the voltages at the L1/C1 and L2/C2 nodes for $R\beta$ = 8.2 kΩ. The oscilloscope is set in X-Y mode at 5 V/div.

7.6 Conclusion

The oscilloscope images confirm the circuit is capable of producing complex behaviors that are characteristic of chaotic phenomena. This chapter is limited in scope to the experimental circuit but the results obtained thus far encourage deeper analysis of the circuit.

References

1. B. Muthuswamy and L.O. Chua, "Simplest chaotic circuit," *International Journal of Bifurcation and Chaos*, vol. 20, no. 5, pp. 1567–1580, (2010).

2. T. Saito, "On a coupled relaxation oscillator," *IEEE Transactions on Circuits and Systems*, vol. 35, no. 9, pp. 1147–1155, (1988).

3. F. Komatsu, H. Torikai and T. Saito, "On a network of chaotic oscillators by intermittently coupled capacitors," *IEEE Transactions on Circuits and Systems I: Fundamental Theory and Applications*, vol. 48, no. 2, pp. 226–232, (2001).

4. Z.L. An, H.S. Zhu, X.R. Li, C.N. Xu, Y.J. Xu and X.W. Li, "Nonidentical Linear Pulse-Coupled Oscillators Model With Application to Time Synchronization in Wireless Sensor Networks," *IEEE Transactions on Industrial Electronics*, vol. 58, no. 6, pp. 2205–2215, (2011).

5. C.W. Wu, "Graph coloring via synchronization of coupled oscillators," *IEEE Transactions on Circuits and Systems I: Fundamental Theory and Applications*, vol. 45, no. 9, pp. 974–978, (1998).

6. R. Kharel, K. Busawon and Z. Ghassemlooy, "Indirect coupled oscillators for keystream generation in secure chaotic communication," *Proceedings of the 48th IEEE Conference on Decision and Control, 2009 held jointly with the 2009 28th Chinese Control Conference. CDC/CCC 2009*, pp. 4099–4104, (2009).

7. Y. Nishio, S. Mori and A. Ushida, "On coupled oscillators networks-For the cellular neural network," *IEEE International Symposium on Circuits and Systems, 1993, ISCAS '93*, pp. 2327–2330 vol. 4, (1993).

8. A. Brambilla, G. Gruosso and G.S. Gajani, "A Probe-Based Harmonic Balance Method to Simulate Coupled Oscillators," *IEEE Transactions on Computer-Aided Design of Integrated Circuits and Systems*, vol. 30, no. 7, pp. 960–971, (2011).

9. S. Moro, Y. Nishio and S. Mori, "Synchronization phenomena in RC oscillators coupled by one resistor," *IEEE International Symposium on Circuits and Systems, 1994, ISCAS '94*, pp. 213–216 vol. 6, (1994).

10. W. Zhang and X.F. Zou, "Synchronization feature of coupled cell-cycle oscillators," *2011 IEEE International Conference on Systems Biology (ISB)*, pp. 190–196, (2011).

11. C.W. Wu, "Synchronization in arrays of coupled nonlinear systems: passivity, circle criterion, and observer design," *IEEE Transactions on Circuits and Systems I: Fundamental Theory and Applications*, vol. 48, no. 10, pp. 1257–1261, (2001).

12. X. Gao, X.W. Liu and S.Q. Shao, "Projective Synchronization in Coupled Fractional Order Chaotic Electronic Oscillators and its Control," *International Conference on Communications, Circuits and Systems 2007, ICCCAS 2007*, pp. 1204–1207, (2007).

13. J. Kurths and C. Zhou, "Noise-enhanced phase synchronization of weakly coupled chaotic oscillators," *Proceedings. 2003 International Conference Physics and Control, 2003*, vol. 2, pp. 353–357, (2003).

14. L.O. Chua and S.M. Kang, "Memristive devices and systems," *Proceedings of the IEEE*, vol. 64, no. 2, pp. 209–223, (1976).

15. L.O. Chua, "Memristor-The missing circuit element," *IEEE Transactions on Circuit Theory*, vol. 18, no. 5, pp. 507–519, (1971).

Chapter 8

Conclusion and Future Work

8.1 Summary

This book has covered a number of systems capable of being used to emulate memristive behavior and the use of these in practical circuits. The focus has been on systems comprised of analog circuitry, rather than micro-processor based and a preference towards simplicity and minimalism in design.

Most notably the emulator circuits lack any memory function; this is the goal of current development for memristor emulator circuits. One possible solution is to implement a sample and hold circuit to hold the voltage corresponding to flux that is often used to set the memristive value in emulators.

Ideally memristors will eventually be available as individual components in a through-hole or surface-mount format and in a variety of values and ratings, in the same way other electronic components can be obtained, this would obviously negate the need for emulator circuits. In the short term, this is not economically viable but would greatly enable and promote intense research into discovering more applications for memristive devices. This reason alone should justify the expense and effort involved in manufacturing standalone memristors, or at least memristor arrays on integrated circuits that can be directly accessed by external analog components.

It is common to see new technologies evolve and be utilized in ways that were never the intention of the originator or quite far from the original expectations of applications for the device. This may well be so with the memristor, so at this stage it is far safer to describe the current research paths being taken to exploit and utilise memristors rather than predict future applications. Whether or not any or all of these research projects will be successful, in terms of products or applications cannot be determined at this stage. There are other competing technologies such as nonvolatile phase change memory being developed by Intel, which uses types of glass and heat to obtain memory functions by altering the atomic structure of the glass and thereby altering its resistivity. This technology is currently aimed at the flash memory market [1] and is likely a competitor only in this field.

The initial production of memristor based products is intended for devices to use in digital computing. The initial applications are flash memory devices, DRAM and SRAM integrated circuits. A number of advantages can be derived from the use of memristor based technology to supplant or complement transistor based systems. Mainly in terms of size, energy consumption and speed, memristors are a desirable option.

The downside, of using memristors to replace transistors, is that the status quo does not really change. Eventually the same issues, such as overheating and limitations in shrinking the size of the component will begin to occur as they are now appearing for transistor based integrated circuits. Furthermore the continued use of digital only computers will stifle the development of practical self-learning devices capable of rudimentary artificial intelligence (AI) and the ability to receive and process complex stimuli.

These next steps in computational ability are more likely to be achieved by the implementation of analog, analog-neural and analog-digital hybrid devices rather than digital only technologies. It is important to move beyond the limitations of iterative binary computing and develop systems capable of large scale parallel processing with analog values.

It can be argued that the next monumental shift in technology and information-processing lies in the development of analog neural computing and a re-examination of the merits of analog computing, especially in the context of current manufacturing standards. The memristor is an ideal component for these types of computing systems as it is an analog device with memory and the ability to switch in the ways digital components are required to do. The memristor can function perfectly as a digital device as shown by the work of Biolek and others [2–5].

8.2 Memory Applications

The first memristor based products expected to be available will be non-volatile random access memory (NVRAM) devices. These can retain the data without power, effectively like the ubiquitous USB thumb drives in common use.

A great deal of work is being undertaken by several companies to develop a new type of non-volatile memory known as resistive random access memory (ReRAM). The basic concept is that a dielectric conducts once an appropriate voltage has been applied. Once the conductance is achieved the resistance of the device can be adjusted by applying another voltage of the corresponding amplitude.

Leon Chua has written that he considers that ReRAM devices to be memristors or at least memristive systems [6]. Further work is being done to develop ReRAM arrays that are self-adapting to new and random data patterns which allow greater energy and storage efficiency [7].

8.3 Low Power Devices and Sensing

The natural consequences of the nano-scale size of memristors are the low power requirements and high density vertically stacked layout in integrated circuits (ICs). This is eminently suitable for mobile devices and implant nano-devices designed to be installed inside living bodies. Research is being undertaken to develop these ICs and suitable applications [8, 9]. The low power requirements are expected to be met by heat or glucose conversion derived from the host body.

Further research is attempting to develop sensing applications using memristors which will be particularly useful in bio-medical fields. These include magnetic sensing to detect flux, magnetic strength, direction and disturbances in a magnetic field. The notable advantages of using memristors for these applications are the small size, low power requirements, compatible with CMOS and excellent sensitivity [10].

Other development of memristor based sensing devices has found the spintronic memristor exhibits thermal fluctuation effects which make it suitable to be used as a basis for a temperature sensing device [11].

8.4 Neuromorphic Applications

It was quickly realized by many that the memristor shares several behavioral characteristics with the synapse and is capable of emulating these [12]. Perhaps Leon Chua's description of the potentials is most succinct:

"Since our brains are made of memristors, the flood gate is now open for commercialization of computers that would compute like human brains, which is totally different from the von Neumann architecture underpinning all digital computers." [13]

Chua further expands this by explaining that the brain is made up of two types of memristors, one used for memory and the other for processing. This implies that the development of circuits that operate in a similar fashion to neural activity is possible with memristors. Already positive and promising results have been obtained with the development of learning circuits by Cantley *et al.* [14, 15]. Another project to implement memristor based neuromorphic devices into semi-autonomous self-learning robotic systems in the MONETA project [16].

A recent trial pitted a CMOS neural network design against a memristor-MOS (MMOST) design; the memristor version showed many improvements notably, its layout was approximately one fifth of that of the CMOS layout. The memristor layout ran at 15 µW, whereas the CMOS version used 55 µW [17].

The similarities between memristors and synapses in neural systems have led to a focus on hybrid neuromorphic computing applications. These use complementary metal-oxide semiconductor (CMOS) devices for the neurons and memristors for the synapses [12]. Jo *et al.* expect the use of memristors in this way will potentially allow the development of nano-scale neural networks to approach the functional density and connectivity of biological systems and enable the networks to operate in a similar fashion [12].

There are a number of obstacles to overcome in the development of practical neuromorphic computers and the proper exploitation of their features. Some of these are outlined by Pershin and Di Ventra in [18], and include the difficulty in the connection layout between neurons; a biological neuron usually has several thousand synapses connected to it. Also the challenges of developing large scale parallel computation are considered.

8.5 Flexible Circuits

Memristors and memristive systems can be created from a number of different elements and materials. The HP version using TiO_2 is simply one method to achieve the memristive effect. Researchers are working to develop memristors suitable for use in flexible circuits [19]. Flexible electronics is not a new field but is one that has become increasingly important as the need to fit circuits into small spaces and folding devices, such as mobile phones, increases. The research has shown that ferroelectric polymers blended with semiconductor materials can exhibit switching and memory effects typical of memristive systems with no significant trade-off or degradation of performance specifications [19].

8.6 Analog Applications

Many of the most interesting applications for memristors can be found in discrete analog circuits. Memristors can be used as programmable potentiometers and in this context greatly enhance the configuration and functionality of standard analog circuits such as comparators, filters and oscillators. These core circuits can be shrunk so many can be laid out on integrated circuits and then the memristors are programmed to set up the circuits as required. It can be expected the circuits can be altered and refined as needed by simply reprogramming the memristors. These types of integrated circuits (ICs) can be expected to appear as an evolution of the Field Programmable Analog Array (FPAA) ICs. A technology that holds much promise but failed to have a significant impact due to its reliance on digital programming and FET based buses.

As demonstrated in this book, memristors are also suitable in the development of chaotic circuits. Once again their nano-scale sizing will allow the easy development of on-chip chaotic circuits that can be used for random number generation, encryption and in some types of neuromorphic computing.

References

1. G. Anthes, "Memristors: pass or fail?" *Communications of the ACM*, vol. 54, no. 3, pp. 22–24, (2011).
2. T. Raja and S. Mourad, "Digital logic implementation in memristor-based crossbars - A Tutorial," *Fifth IEEE International Symposium on Electronic Design, Test and Application, 2010. DELTA '10*, pp. 303–309, (2010).
3. G.S. Rose, J Rajendran, H. Manem, R. Karri and R.E. Pino, "Leveraging memristive systems in the construction of digital logic circuits," *Proceedings of the IEEE*, vol. 100, no. 6, pp. 2033–2049, (2012).
4. D. Biolek, V. Biolkova and Z. Kolka, "Mutators simulating memcapacitors and meminductors," *2010 IEEE Asia Pacific Conference on Circuits and Systems (APCCAS)*, pp. 800–803, (2010).
5. K. Bickerstaff and E.E. Swartzlander, "Memristor-based arithmetic," *2010 Conference Record of the Forty Fourth Asilomar Conference on Signals, Systems and Computers (ASILOMAR)*, pp. 1173–1177, (2010).
6. L.O. Chua, "Resistance switching memories are memristors", *Applied Physics A*, vol. 102, no. 4, pp. 765–783, (2011).
7. S. Shin, K. Kim and S.M. Kang, "Analysis of passive memristive devices array: data-dependent statistical model and self-adaptable sense resistance for RRAMs," *Proceedings of the IEEE*, vol. 100, no. 6, pp. 2021–2032, (2012).
8. P.F. Chiu, M.F. Chang, C.W. Wu, C.H. Chuang, S.S. Sheu, Y.S. Chen and M.J. Tsai, "Low store energy, Low VDDmin, 8T2R nonvolatile latch and SRAM with vertical-stacked resistive memory

(Memristor) devices for low power mobile applications," *IEEE Journal of Solid-State Circuits*, vol. 47, no. 6, pp. 1483–1496, (2012).

9. J. Rajendran, H. Manem, R. Karri and G.S. Rose, "An energy-efficient memristive threshold logic circuit," *IEEE Transactions on Computers*, vol. 61, no. 4, pp. 474–487, (2012).

10. Y. Chen, X.B. Wang, Z.Y. Sun and H. Li, "The application of spintronic devices in magnetic bio-sensing," *2010 2nd Asia Symposium on Quality Electronic Design (ASQED)*, pp. 230–234, (2010).

11. X.Y. Bi, C. Zhang, H. Li, Y. Chen and R.E. Pino, "Spintronic memristor based temperature sensor design with CMOS current reference," *Design, Automation & Test in Europe Conference & Exhibition (DATE), 2012*, pp. 1301–1306, (2012).

12. S.H. Jo, T. Chang, I. Ebong, B. B. Bhadviya, P. Mazumder and W. Lu, "Nanoscale memristor device as synapse in neuromorphic systems," *Nano Letters*, vol. 10, no. 4, pp. 1297–1301, (2010).

13. "HP labs discovery holds potential to fundamentally change computer system design "Memristor" could enable computation on memory chips," *HP Newsroom News release*, 8 April 2010. Available from:
http://www.hp.com/hpinfo/newsroom/press/2010/100408xa.html

14. K.D. Cantley, A. Subramaniam, H.J. Stiegler, R.A. Chapman and E.M. Vogel, "Neural learning circuits utilizing nano-crystalline silicon transistors and memristors," *IEEE Transactions on Neural Networks and Learning Systems*, vol. 23, no. 4, pp. 565–573, (2012).

15. K.D. Cantley, A. Subramaniam, H.J. Stiegler, R.A. Chapman and E.M. Vogel, "Hebbian learning in spiking neural networks with nano-crystalline silicon TFTs and memristive synapses," *IEEE Transactions on Nanotechnology*, vol. 10, no. 5, pp. 1066–1073, (2011).

16. M. Versace and B. Chandler, "The brain of a new machine," *IEEE Spectrum*, vol. 47, no. 12, pp. 30–37, December 2010.

17. I. Ebong and P. Mazumder, "CMOS and memristor-based neural network design for position detection," *Proceedings of the IEEE*, vol. 100, no. 6, pp. 2050–2060, (2012).

18. Y.V. Pershin and M. Di Ventra, "Neuromorphic, digital, and quantum computation with memory circuit elements," *Proceedings of the IEEE*, vol. 100, no. 6, pp. 2071–2080, (2012).

19. N. Gergel-Hackett, J.L. Tedesco and C.A. Richter, "Memristors with flexible electronic applications," *Proceedings of the IEEE*, vol. 100, no. 6, pp. 1971–1978, (2012).

Index